Tapestry Lawns

Tapestry Lawns

Freed from Grass and Full of Flowers

Lionel Smith

CRC Press
Taylor & Francis Group
Boca Raton London New York

CRC Press is an imprint of the
Taylor & Francis Group, an **informa** business

CRC Press
Taylor & Francis Group
6000 Broken Sound Parkway NW, Suite 300
Boca Raton, FL 33487-2742

© 2019 by Taylor & Francis Group, LLC
CRC Press is an imprint of Taylor & Francis Group, an Informa business

No claim to original U.S. Government works

Printed on acid-free paper

International Standard Book Number-13: 978-0-367-20747-2 (Hardback)
978-0-367-14403-6 (Paperback)

Library of Congress Cataloging-in-Publication Data

Names: Smith, Lionel (Lionel Stephen), 1965- author.
Title: Tapestry lawns : freed from grass and full of flowers / author: Lionel Smith.
Description: Boca Raton, FL : CRC Press, Taylor & Francis Group, 2019. | Includes bibliographical references.
Identifiers: LCCN 2019016587 | ISBN 9780367144036 (pbk. : alk. paper) | ISBN 9780367207472 (hardback : alk. paper)
Subjects: LCSH: Lawns. | Gardening. | Ornamental horticulture.
Classification: LCC SB433 .S6162 2019 | DDC 635--dc23
LC record available at https://lccn.loc.gov/2019016587

Visit the Taylor & Francis Web site at
http://www.taylorandfrancis.com

and the CRC Press Web site at
http://www.crcpress.com

Contents

Preface

Like all things an idea has a beginning. For me it was a question that started things off, and like so many other questions formed at a young and curious age, it started with 'Why?'

As I recall, it was when I was 11 years old, and my family had just returned to our home in Bedford from a family holiday on the then sun-drenched East Coast of England, where we had experienced a plague of ladybirds that had followed an earlier plague of greenfly. It was 1976, the year of the Great Drought. Britain was unusually sweltering in relentless weeks of heat and sunshine, and there had been no rain to speak of; the reservoirs were running dry. We had a brick in the toilet cistern and only flushed when it was an absolute necessity, had discovered the meaning and value of 'grey water', and had watched the ground in the garden split and crack like only Oxford clay can. For an outdoor sort of boy of 11, the weather was glorious!

I remember us arriving home and looking out of the window as the car came to a halt on the drive, and seeing that during our holiday virtually all the plants in the front garden borders had succumbed in some way to the heat and lack of water. All the grasses in the front lawn had shrivelled and browned, but notably not the so-called weeds. I particularly remember the ribwort plantains were in full flower, like firework sparklers frozen in time, their pale pollen almost exploding on wiry stems along with the less demonstrative but very robust greater plantain.

There were the delicate globes of the odd dandelion seed head and yellow hawksbeard and cat's-ear in bloom, along with daisies, both red-and-white clovers, orange-tinted bird's-foot trefoil (that all the kids called 'eggs & bacon'), creeping cinquefoil ('finger-leaf'), purple selfheal and brilliant yellow buttercups. I had never seen the lawn so devoid of grass, so floriferous, and to my eyes so incredibly beautiful. I remember my father saying, 'Lawn needs a haircut'.

I distinctly recall I protested, really protested, but my arguments fell on deaf ears. It was part of the deal for my pocket money, I would be the destroyer of the most beautiful 'lawn' I had ever seen.

FIGURE 1   In a young boy's mind – living sparklers in the lawn. Ribwort plantain (*Plantago lanceolata*) in bloom. Not especially useful in tapestry lawns but inspirational nonetheless.

Before I did the deed with our very heavy hand mower, I lay flat on the ground for a while to get a mouse-eye view of the spires of flowers and watch the myriad of bees, hoverflies, orange ants, curious-looking beetles and the occasional butterfly; they seemed to like it there as much as I did. It was as I gave a crew cut to the little grass-free flower meadow that our front lawn had become that the questions arose 'Why do lawns have to have grass?' and 'What would happen if the grass wasn't there?'

Thirty years later I found myself on the road to answering some of those questions, and much of what I have learned on the way I have attempted to share here. I have endeavoured to include both some of the thinking behind the 'why' the format is possible, and the practicalities of the 'how' to make it possible, so that others might develop an understanding of the major influences and issues involved and create or manage their own tapestry lawns if they should choose to do so.

As with any new approach to almost any topic, the acquisition of knowledge and its subsequent development and application take time. As any gardener and landscape manager will know, developments within horticulture rely ultimately on the slow progress of seasons and years. Tapestry lawns are, therefore, not a done-and-dusted development; there is still much to learn.

Since the research, development and planting of tapestry lawns has occurred within the British Isles, it has specifically used plants that are essentially at home throughout these islands, or those that are climatically suitable and available within current British ornamental horticulture. These plants may not be ideal in other parts of the globe, and the format constituents may have to be amended accordingly if attempted elsewhere.

Lawns, like all living garden features, change over time and are dependent on the ongoing management and maintenance provided by those who look after them. Circumstances and priorities change, and not everything necessarily goes right the first time. Alas, not all the T-lawns referred to in this book remain to be seen. However, it is my hope that the questions raised by a curious 11-year-old boy and the contents of this book will inspire gardeners and landscape managers to look anew at a venerable garden feature that could do with the cobwebs kicking out of it.

Author's note: 'Forb' is an uncommon word that will crop up quite a bit in this book, so it is worth knowing what it means. A forb is a dicotyledonous (has two baby leaves when it germinates rather than the single leaf that grasses have), herbaceous (without a true woody stem) perennial (can live longer than two growing seasons). Specifically, forbs are not grasses, and they may be herbaceous (die down in a usual winter), semi-evergreen (partially die down in a usual winter) or evergreen (do not generally die in usual winters). 'Herb' is a word sometimes used synonymously, especially in North America, and particularly for herbaceous plants, but in common parlance in the UK 'herb' is generally understood to mean a culinary variety of plant.

Acknowledgements

My thanks go to the University of Reading for the little hiccup that let me start the research first and worry about complete funding later, and to Myerscough College for letting me continue planting and monitoring the lawns. My gratitude to Professors Mark Fellowes and Paul Hadley for all their knowledge and invaluable support, and Valerie Jasper for practical horticultural assistance throughout the research. My sincere thanks to the Royal Horticultural Society, particularly Leigh Hunt, and both Roger Williams and John David, without whom this new approach to lawns would not have been investigated to the depth it has. I am grateful to the horticultural curiosity of Sue Allen and Gillie Westwood, and the constancy of Pat Adams of the Garden Centre Association's Dick Allen Scholarship Fund, and to the Finnis-Scott Foundation for supporting my seemingly odd-ball idea.

Additionally, my appreciation to the Stanley Smith Horticultural Trust for helping me show the lawn format to the wider public at the 100th RHS Chelsea Flower Show, and the Gilchrist Educational Trust for giving me the resources to finish the final thesis. To Lara Hurley for her artistic skills, the wonderful doctors and nurses of the Haematology and Oncology ward of the Victoria Hospital, Blackpool, and not least to my good friend of many years, Simon Bass, for chipping in at the start to help make it all possible and get me out of the house.

Author

Lionel Smith was given a small plot in the back garden by his parents at age seven and has been fascinated by plants ever since. He was awarded his PhD in 2014 for his research into tapestry lawns and currently lives beside the Irish Sea.

About the Book

Swathes of the human world are covered in ornamental grass lawns; they are the singlemost commonly encountered horticultural feature on the planet. Unfortunately, they are now often viewed as resource-draining green deserts due to the lack of plant and animal diversity, the need for frequent mowing and watering, and addition of lawn-greening products to keep them looking at their best. It is a venerable horticultural feature that is essentially frozen in time, and with few alternatives to whet the appetite, the lawn has languished in its current grass-only format for decades. Until now.

Tapestry lawns are a new, practically researched and timely development of the ornamental lawn format that integrates both horticultural practice and ecological science and redetermines what a lawn can be.

Mown barely a handful of times a year and with no need for fertilisers or scarifying, tapestry lawns are substantially richer in their diversity of plant and animal life compared to traditional grass-only lawns, and see the return of flowers and colour to a format from which they are usually purposefully excluded.

This book traces the changes in the lawn format from its origins to the modern day and offers information on the how and why the tapestry lawn construct is now achievable. It provides guidance on how to create and maintain a tapestry lawn of your own and the potential benefits for wildlife that can follow. A set of useful tapestry lawn plants is also included.

If you have ever thought about mowing your lawn a lot less and making it much more colourful and wildlife friendly, then this book will inform and guide you and have you rethinking what a lawn really can be.

1 'Lawn' a Word That Is Strangling Itself

You might not think it, but the word 'lawn' is rather problematic. It really is. It currently conjures a very particular and somewhat constrained image of an area of thickly textured, soft, verdant grass, uniformly levelled and striped with lines laid by a finely bladed mower. The kind of lawn that 'KEEP OFF THE GRASS' signs are especially made for (Figure 1.1).

This is perhaps a little disheartening if you happen to have a typically average British-style garden lawn with its mix of different grasses and varying heights (Figure 1.2), with dandelions, buttercups and other plants that weren't originally invited (Figure 1.3), and it's quite an unfortunate image since lawns are and have always been so much more than just the restricted selection of grasses we are familiar with today.

To better understand what the lawn has been and can be, it is worth taking a brief look back in time at its origins and history. It will help clarify a few things.

Let's start with that problematic word. The word 'lawn' itself is regarded as having a number of potential origins from around the fourteenth century; perhaps 'launde' in Old French, 'laund' in Old English or possibly 'lawnd' in Middle English; each of these being used to describe an open and pastured glade especially within woodland [1]. Another suggested source is the name of the French town of Laon, which is pronounced rather like the English word 'long' but without the 'g'. It was renowned for producing types of linen cloth made using carded or combed linen in a simple plain weave that produced particularly fine silky fabrics known as 'laune lynen' and commonly referred to as 'lawn' [2]. Exactly how a French town's silky fine fabric may have given up its name to a collection of mixed cut grasses and forbs remains hidden in the murky mists of time and linguistics, or it just might not be relevant at all.

Whatever the origins of the word we currently use today, keeping grasses short for some aesthetic purpose is a practice older than the word we now use for it. There is a niggling suggestion that landscape lawns were in existence in Italy during the era of Imperial Rome, but here we have another problem with words again, since the English translation of the relevant Latin text that is thought to mention a lawn also goes on to make mention of a Roman tennis court [3]. With tennis thought to have its origins in twelfth-century France, it seems unlikely that there were actual tennis courts in ancient Italy, perhaps 'ball court' might have been a better translation. It also seems likely that the translator was applying modern words to ancient practices; the Latin word 'Pratum' meaning 'prairie' or 'meadow' being translated into 'lawn' to suit the context of a wealthy aristocrat's villa's garden. Other translators have used 'level ground' instead for the same text, so perhaps there were no actual frequently mown lawns as we now understand them in the ancient roman world either, perhaps 'grassy ground' might be a better descriptor if the few period garden frescoes are anything to go by. If there were cultivated lawns, and it's a notable 'if' considering that the Mediterranean-type climate is not usually kind to moisture-loving lawn grasses and the period between 250BC and 400AD is thought to have been unusually warm; they were possibly seasonal features or likely to have been necessarily restricted to areas where an irrigating supply of water was available throughout the year, whichever it might possibly have been they weren't clearly mentioned or shown and apparently didn't catch on.

Head farther north and west in Europe to the temperate oceanic climate of northern France and the British Isles (Figure 1.4), and the irrigation of grasses is much less challenging due to the relatively mild temperatures and substantial amount of moisture carried in from the Atlantic Ocean by weather systems associated with the Gulf Stream ocean current.

FIGURE 1.1   Weed-free, thickly textured, verdant green and showing the stripes of the mower. What many a gardener and groundsman would regard as the 'ideal' lawn.

FIGURE 1.2   The more commonly encountered British-style lawn is a mix of different grasses and uninvited weeds.

There are records of English King Henry II's (1113–1189 AD) palace at Clarendon, Wiltshire, containing gardens which were said to 'boast a wealth of lawns' [4], although that word still causes problems, and they too may have been pastured meadows. However, this potentially moves the provenance of purposefully keeping grasses low and recognising it as a specific feature to at least the early twelfth century and very probably earlier, although by what method the grasses were kept

FIGURE 1.3   Even well-managed lawns rarely keep the sharp-cut look for long. Despite the evidence presented to their eyes, the mental image conjured by the word 'lawn' has many people envisioning uninterrupted flat surfaces.

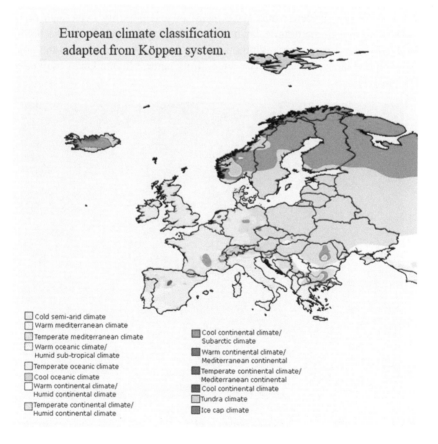

FIGURE 1.4   Köppen–Geiger climate classification zones of Europe. The temperate oceanic climate of NW Europe (shown in light green) fosters the luxuriant growth of certain grasses.

short remains unclear; it seems likely that it was grazing animals or possibly scythes collecting hay or perhaps a mixture of both. By whatever method the grass was made or kept short, it does indicate that the resulting lawn was an identifiable landscape feature worthy of a name, and thus lawns of some sort may have been around for at least a millennium.

Also, in Britain in 1259 AD during the reign of Henry III, it was recorded that the gardens of the Palace of Westminster were levelled with a roller, turf laid and later mown [5]. This now does sound very familiar and is clearly typical lawn-making behaviour. Additionally, in the same century the creation of what is quite clearly a lawn was summarised by the Swabian nobleman Albert Magnus, Count of Bollstadt, in the chapter 'On the Planting of Pleasure Gardens' in his thesis De Vegetabilibus et Plantis (On Vegetables and Plants) [6].

Along with a use as a feature in pleasure gardens there are also sports being recorded as being played on grass lawns, with Southampton Old Bowling Green recorded as hosting continuous lawn bowl games since 1299 AD, and as any enthusiast will tell you, lawn bowls requires a very well managed type of lawn indeed, not just any old grass will do; it must provide a fine and even surface on which to play. We may infer that even early on in its history there were different types of lawns, with different uses that were composed of different types of grasses accordingly.

The plant constituents of these early grass lawns are not clearly recorded, but when they are referred to it is apparent that grass lawns were considered quite distinct from the other types of lawns that also existed at the time. Other lawns? Indeed, yes, there were other lawns.

OTHER LAWNS

The informed or curious gardener may have heard of and possibly even seen or touched and inhaled the fragrance of a chamomile lawn (Figure 1.5), and maybe if in the habit of visiting ancient castle gardens or seed company premises may have even encountered the much rarer and just as fragrant thyme lawn. Occasionally even entirely natural thyme lawns may be encountered (Figures 1.6 and 1.7).

FIGURE 1.5   A chamomile lawn (*Chamaemelum nobile*). As with most lawns, coverage can be variable, and unwanted plants such as grasses, daisies and clovers can appear. The keepers of this lawn report they spend 'a lot of time weeding it'. In much larger and frequently used chamomile lawns such as that at Buckingham Palace the chamomile is mixed with lawn grasses and is not managed as a pure species lawn at all.

FIGURE 1.6   A natural thyme lawn in flower in the dunes, St. Anne's-on-Sea, Lancashire. With the day-flying slender scotch burnet moth (*Zygaena loti*) taking nectar from the flowers. Natural thyme lawns tend to be small and include a mix of other hardy wild plants.

FIGURE 1.7   A natural dune mixed-species lawn on almost pure mineral sand with the forbs *Thymus praecox* (mother of thyme), *Centaurium erythraea* (common centaury), *Lotus corniculatus* (bird's foot trefoil) and *Hypochaeris radicata* (cat's ear) amid the sparse grasses and dry moss.

These fragrant and floral lawns have a pedigree as old as grass lawns, but the key features here are not specifically a green sward or a playable surface, but rather distinctly the focus is on scented leaves and seasonal flowers. Both *Chamaemelum nobile* (noble or Roman chamomile) and *Thymus praecox* (mother of thyme) are strongly aromatic, most especially when the leaves are bruised or crushed underfoot.

It is the kind of fragrance that a bit of treading can release rather well. 'Treading the lawn' was a practice well-understood in a time when using strewing herbs such as chamomile, mint,

FIGURE 1.8   The edge of a chamomile lawn. Indicative of how mowing and treading can influence plant growth and behaviour. Where the mower has not reached and the lawn is untrodden the growth is upwards towards flowering rather than outwards and ground-covering. The non-flowering and creeping cultivar clone 'Treneague' can be used to overcome this, especially if trodden on from time to time.

lavender, pennyroyal, meadow sweet and marjoram on floors of houses was undertaken in all the very best households (King Charles II appointed his own royal herb strewer). The venerable playwright William Shakespeare even makes mention of the custom known as 'treading the chamomile' in King Henry IV, Part 1 'Though the chamomile, the more it is trodden upon, the faster it grows' (Figure 1.8).

Treading, walking, laying and even dancing on plants of all sorts doesn't seem to have particularly bothered our medieval forebears, possibly because of all that strewing going on. Tapestries and paintings of the period from across Europe show the noble and wealthy classes taking their ease in enclosed gardens, both on and within grassed areas embellished with flowering and scented plants of all sorts (Figure 1.9). It is not just the daisies that are shown in some medieval lawns but irises, lilies, aquilegias, lily of the valley, borage, mints, campions, poppies, hellebores, cowslips, primroses, wild strawberries, violets and daffodils to name but a few. Such richly enamelled grasslands are generally termed 'flowering meads' and are skilfully cultivated if somewhat romantic imitations of European forb rich meadows (Figure 1.10).

At the time of this writing, flowering meads of a sort are having something of a renaissance, albeit with some component plant species now globally sourced and in a less danced-upon format. The use of 'wildflower' meadows and mixed grasses and perennials in what is often described as 'prairie planting' is something of a relatively recent trend among the horticultural cognoscenti, although perhaps rather like genuine meadows and prairies it is a format best appreciated on a scale not frequently found in average suburban gardens.

Almost inevitably as the centuries passed there have been a few more refinements and types of lawn to appear. In seventeenth-century France, a very refined and extensive type of lawn known as tapis vert (green carpet) emerged. Cultivated at the Palace of Versailles on the 'Allée royale', the royal lawn at Versailles was reputed to be initially composed of only one species of fine grass to distinguish it from the coarser and more practical pastureland species non-royals would use, a French king not having to be particularly practical about these things; at least that's the tale. Interestingly in 1709 a gardening guide by A. J. Dezallier d'Argenville called

FIGURE 1.9 A medieval flower lawn depicted in the Hours of Henry VIII. The month of April. ca. 1500. MS H.8 (fol. 2v). (Courtesy of Morgan Library, New York.)

FIGURE 1.10 A richly enamelled flowering mead. Hortus Conclusus ca. 1410. (Courtesy of Städelsches Kunstinstitut, Frankfurt, Germany.)

La Theorie et la Pratique du Jardinage (*The Theory and Practice of Gardening*) [7] indicates that another lowlier form of tapis vert was also created using a combination of fine low-growing meadow grasses, catmint (*Nepeta cataria*), vetches (*Vicia* sp.) and clovers (*Trifolium* sp.). We can only speculate on what eighteenth-century French cats were up to on these lawns, but once again, flowers and a heady scent are features of the lawn and the constituents are intentionally not restricted to grasses alone.

Across the English Channel, mixed-species lawns were also a topic of interest. In his 1806 book *A Treatise on Forming, Improving, and Managing Country Residences*, the Scotsman John Claudius

Loudon, who has been hailed as the 'father of the English garden' [8], made the comment 'In almost every case, where lawn is not fed by sheep, it should not be formed of mere grasses; which require continual mowing, and present one dull, vapid, surface of uniform green. They should be composed of primroses, violets, common and garden daisy, camomile, graphallium, doicum, hieracium pilosella and especially white clover' [9].

The inclusion of forbs such as clovers in lawns is hardly surprising when you think about it. Clovers have lawn-colour enhancing rich-green leaves and are in a plant group known as legumes that make important symbiotic associations with soil-dwelling bacteria called *Rhizobium* that form nodules on their roots. There are often other bacteria in those nodules too, but we are not yet sure exactly what they are up to. The rhizobia bacteria in the nodules extract nitrogen from the air (there is air in soil; it's an essential component without which most plant roots suffocate and die) and when in due course the roots degrade and rot away, frequently as a response to defoliation such as mowing or being grazed, traces of organic nitrogen are left behind. These traces are quickly scavenged by microorganisms and neighbouring plants since bio-available nitrogen is the rarest of the essential nutrients used by plants. In effect, most legumes are non-synthetic sources of regular small amounts of nitrogen fertiliser; it's one of the reasons why some clovers can be quite vigorous and survive in relatively nutrient poor soils. Anyone who has ever purchased a lawn rejuvenation product may well have spotted that nitrogen is by far the most significant ingredient and may also have noticed that clovers don't respond well to the additional fertiliser. It tends to act rather like an overdose, discouraging nodulation and inhibiting nitrogen fixation, but since it is aimed at invigorating the grass rather than the clover, most people don't seem to mind.

Historically, lawn seed mixes in the nineteenth and early twentieth centuries also included white clover (*Trifolium repens*) at usually between 5% and 10%. This was before the general use of synthetic fertilisers on lawns, and such a combination was considered to produce a greener and superior type lawn. The advent of the first broadleaf herbicide 2,4-dichlorophenoxyacetic acid (2,4-D) in the 1940s largely put an end to not only clover in treated lawns but to all but the most pernicious of broadleaf lawn weeds, and thus the very recent and subsequently problematic image of the suburban weed-free lawn was born. The weed-free and grass-only lawn as a common garden feature is really a very new development indeed, unless perhaps, you happen to be historically royal.

Today there is something of a commercial move to reintroduce a recently selected form of white clover to lawns. Termed 'microclover' it is a particularly small-leaved variety that shows a degree of drought tolerance, tends to creep extensively rather than form clumps and can be used to enhance the overall greenness of the lawn since it tends to be a deeper shade of green than most grasses.

So why not a clover-only lawn? Certainly, pure clover lawns exist in private gardens in both Europe and north America, but like thyme lawns they are rare and still something of a novelty (Figure 1.11).

An explanation for this is that white clover (*Trifolium repens*) can lose its foliage and become a brown mush in cold winters, and this is particularly noticeable when grown as a single species; brown mush lawns not being particularly popular. Also *T. repens* does not grow well for any extended period as a simple monoculture. The exact reasons behind this are still not well understood, but almost all pure white clover lawns tend to degrade to a coverage of around 60% or so over time and need continuous reseeding to maintain the good ground cover expected of a lawn.

Another very common lawn forb that has been used as a single species in bright and partially shaded lawns is selfheal (*Prunella vulgaris*), but it too benefits from being annually reseeded to provide continuous coverage, and it too can be noticeably cold sensitive when grown as a monoculture (Figure 1.12). Ornamental monocultures in general tend to require regular topping-up if they are to be maintained for anything but the shortest period.

Single species forb lawns have been popping up on and off all over the place, especially where temperate grasses tend to struggle or are inappropriate. They include quite a variety of surprising plant species. In semi-tropical regions such as Florida, a type of South American perennial creeping peanut (*Arachis glabrata*) is increasingly encountered as a floriferous grass-lawn alternative,

FIGURE 1.11   Not a true managed clover lawn, but here white clover predominates in a lawn-like sward.

FIGURE 1.12   A selfheal (*Prunella vulgaris*) lawn in spring, Oregon, USA. (Prichard, H.J., Selfheal lawn, Personal Communication, 2010.)

the carpet gazania (*Dymondia margaretae*) has overtaken quite a few lawns in southern Africa, and California hosts a fair few drought-tolerant yarrow (*Achillea millefolium*)-based lawns.

At the other end of the spectrum in cool, shady and moist positions, we leave the flowering vascular plants behind entirely, with lawns composed entirely of mosses, as can be found in Japanese garden horticulture and elsewhere where the conditions favour bryophytes (Figure 1.13).

FIGURE 1.13   One of two adjacent moss lawns at Dunham Massey, Cheshire, UK.

There is a growing list of lawn-tolerant forbs that can be more suited to the location than grasses, and to the purpose of the individual lawns themselves. Should you ever embark upon a global lawn quest you may encounter single species or occasionally dual species forb lawns composed of such plants as yarrow (*A. millefolium*), corsican mint (*Mentha requienii*), smooth rupturewort (*Herniaria glabra*), lippia/frogfruit/turkey tangle/matchweed (*Phyla nodiflora*), strawberry clover (*Trifolium fragiferum*), carpet gazania (*Dymondia margaretae*), alpine water fern/Antarctic hard fern (*Blechnum penna-marina*), bird's foot trefoil (*Lotus corniculatus*), pennyroyal (*Mentha pulegium*), pratia/bluestar creeper (*Lobelia pedunculata*), pratia/panakenake/lawn Lobelia (*Lobelia angulata*), pearlwort/Irish or Scottish moss (*Sagina subulata*), lawn pennywort (*Hydrocotyle sibthorpioides*), tideturf/swampweed (*Selliera radicans*), capeweed (*Arctotheca calendula*), mercury bay weed (*Dichondra repens*), beach strawberry (*Fragaria chiloensis*), barren strawberry (*Waldsteinia ternata*), creeping mazus (*Mazus reptans*), brass buttons/cotula, piri-piri/sheep's burr (*Acaena inermis*) and sedges (*Carex* spp). This list is by no means definitive and continues to grow as gardeners and landscapers adapt and contend with their local climates and conditions and seek for different outcomes beyond that provided by or even possible with a lawn composed purely of temperate European grasses.

Most of the forbs that are currently used for making T-lawns tend to be listed in horticultural books as low ground covers, lawn replacements or lawn alternatives; as if somehow the low cover provided by them is in some way not a real lawn. For a garden feature that is without a doubt the most superlative example of ground cover itself, this seems rather odd and is additionally perplexing for those of us with an interest in where a plant comes from, since more often than not they are found in garden centres and plant nurseries labelled as 'alpines', primarily because they are short in stature rather than because of any mountainous origin. Occasionally this classification can be very misleading. For example, it is possible to find in the alpine section of many British garden centres a charming little plant known as 'blue moneywort' or 'false pimpernel' (*Lindernia grandiflora*) (Figure 1.14). It is native to the southeast of the United States and grows wild in the great alpine state of Florida.

However, this does lead to yet another type of lawn worth a mention. A visit to many high mountainous regions of the world, particularly in early spring, will reveal yet another lawn form, and one that is generally thought of as naturally to semi-naturally occurring. Unsurprisingly it is known as an 'alpine lawn' (Figure 1.15). However, the natural alpine lawn is somewhat poorly characterised. It can be a type of vegetation-defined band around mountains dependent on climate and on season

FIGURE 1.14 Often found in the 'Alpine' section of garden centres, *Lindernia grandiflora* (blue money-wort) is prostrate and mat forming but is native to Florida and the SE USA.

FIGURE 1.15 A natural Alpine lawn in early spring, Slovenia. (Griffiths, K., Alpine lawn, Hochobir Mountain, Slovenia, Personal Communication, 2006.)

but is generally a feature that tends to occur where scree becomes less dominant on a high moun-
tainside. This is a relatively more stable area where a thin soil can develop and it can support rocky
pasture and rather stunning alpine flower meadows.

The mixture of plants found there tends to include those that are small and compact and have dispro-
portionately large flowers to attract pollinators, and they have enchanted gardeners for centuries. They
are best viewed early in the growing season since the grasses that also inhabit this region can, given the
chance, grow to be taller than the generally dwarf alpines. The overall effect can be quite enchanting
before the little plants are eventually overwhelmed and the lawn becomes more akin to an alpine meadow.

Inevitably this type of lawn has inspired gardeners to try something similar at home but has not met
with as much success as might be anticipated. Planting alpines in grasses, even fine low-growing types
of grass, eventually sees the alpines swamped [10], and when grasses are excluded and only creeping
alpines used, the largest and most vigorous of them tend to eventually dominate without continuous
intervention. The overall effect is rather different from the generally monotone and evenly cut surface
we might expect of a mown grass lawn, since each species has its own growth form and leaf colouration
and there may be hummocks and mounds and different hues running throughout the alpine-style lawn.

There is also nothing that says that an alpine lawn must be horizontal either or restricted to low-
growing or creeping forbs. It can be sloped or shaped to match the terrain but almost invariably
tends to be dotted with frequent boulders and rocks. Enthusiasts may also embellish it with low-
growing alpine shrubs and taller perennials such as *Aquilegia* sp. and *Iris* sp., but here perhaps we
begin to move away from the essentially low aspect format we familiarly associate with lawns and
begin to stray into the realms of rock gardening or alpine-style planting (Figure 1.16).

FIGURE 1.16   An alpine-style rock garden showing the variety of heights, shapes and forms found in alpine
plants. Although the plants can spread and mix with their neighbours, the rocky outcrops, sharp drainage and
use of woody plants make it unsuitable for a mown lawn.

A NOTE ON ALPINES

Since an alpine lawn can be enchanting it is a wonder that we don't see more of them, but true alpine plants have evolved to survive high alpine conditions above the treeline and are not found in the relatively warm and consistently moist lowland climates of the British Isles and northern western Europe where lush grass lawns are the norm.

The high intensity of ultraviolet radiation at mountainous heights influences leaf and plant growth patterns, plant size, drought tolerance and even the incidence and severity of pests and diseases. True alpine plants are adapted to extreme fluctuations of temperature with optimum rates of respiration and photosynthesis at lower temperatures than those experienced at lower altitudes, and they are usually well suited to nutrient-limited environments.

There are lower altitude forms of many alpine plants, and some species have a surprisingly broad range. It is these that tend to be a better option in lowland situations, but even they generally prefer sharp drainage and strong light. Importantly, seasonal changes in alpine regions lock up liquid water in the winter as blankets of snow and only release it in spring, reducing the likelihood of elemental exposure and constantly winter-wet soil; which as many experienced gardeners will tell you is (along with slimy molluscs) the bane of most alpine plants grown in temperate lowland gardens. The essentially low altitude, nutrient-rich, non-alpine conditions found in most suburban gardens tends to exclude lawns made with true alpines as a viable proposition for all but the most determined rock-owning enthusiast.

Alpine lawns remain rarely encountered outside of alpine regions, but if you head to the antipodes, you may find the low, creeping, alpine-like forbs of New Zealand being used in novel ways. Some of their own native plants have inevitably crept their way into the British-style lawns common across the similarly maritime islands, initially as weeds of course, but that is not always the viewpoint now (Figure 1.17).

Back in 1913, in Dunedin in the far south of the South Island, to the irritation of the groundsman of the Caledonian Bowls Club, the native 'weed' species *Leptinella dioica* (Syn. *Cotula*) had invaded the carefully managed bowling lawns. This was before the invention of selective herbicides, and an attempt at its removal by scarifying had the opposite effect. Over time the low-growing and creeping weed came to dominate the lawn, and with some surprise, the players noticed how the

FIGURE 1.17 *Lobelia angulata* growing in the fine turf lawns of Government Gardens, Rotorua, New Zealand. Once considered a weed this plant is now often called New Zealand's favourite native ground cover plant.

weedy surface played an apparently superior and faster game. In 1930, grass was usurped as the only appropriate surface for lawn bowls as the Caledonian Bowls Club became the first bowls club in the world to replace grass entirely.

Another member of the same genus *Leptinella traillii* sometimes attributed as *Leptinella maniototo* is now also used in combination with *L. dioica* in the creation of fast bowling lawns, with modern techniques such as hydro-seeding being used to impregnate the bowling lawn surface with *Leptinella* bulbils.

At the northern end of the South Island some other low-growing native forbs (and some European exotics) have been used to create lawns more appropriate to the local conditions and in response to an increasing awareness of the value of New Zealand's native flora.

The primary attraction of these New Zealand lawns with alpine-like creeping native plants is that they do not require any mowing at all and are successfully marketed as such, which raises yet another question about lawns: do they also have to be mown to be real lawns and if so how often?

Mowing artificially reduces the height of a lawn, but there is another circumstance other than being mown that can restrict grasses from becoming the dominant vegetation and can produce something very lawn-like. 'Machair' grassland uniquely occurs over sandy, calcareous soils of coastal sand-plains in the dune systems of north-western Scotland and Ireland, particularly around the Hebridean Islands (Figure 1.18).

The islands have a cool climate, are subject to the full force of the Atlantic's westerly winds and receive some of the most regular and copious rainfall in the UK. The very nutrient-poor and often well-drained soils in this region are complex but generally have a low organic matter content of rarely more than 10% and can be up to 90% calcium carbonate, predominantly from ocean-crushed seashells.

Humans have long interacted with and managed this unique environment with crop-rotations, including long fallow periods and winter grazing for sheep and cattle; the nutrient-poor soil being enriched and stabilised with dung from the animals and the spreading of lots of soil-binding seaweed brought in by the Atlantic winter gales. These conditions and their historical management

FIGURE 1.18   Floristically diverse seaside machair, Outer Hebrides, Scotland.

have resulted in a unique and floristically diverse plant community with up to 45 plant species per square metre and a correspondingly rich insect community (Figures 1.19 and 1.20).

With their advantage for swift and tall growth compromised by seasonal grazing and nutrient limitation, the subsequently less-dominant grasses of the machair are indicative of how useful repeated defoliation and a low nutrient environment can be in fostering overall plant diversity.

FIGURE 1.19   Sea-side machair plants blooming in what for all intents and purposes can appear to be an enormous flower-spangled lawn.

FIGURE 1.20   A closer view of machair turf. Even though the Outer Hebrides receive substantial and regular amounts of rainfall, grasses are disadvantaged by the winter grazing and low nutrient availability letting forbs proliferate, although they too can be dwarfed by the conditions. Almost all the plant species found there can exist in the tapestry lawn format.

Hopefully by now you will be on your way to joining me in thinking that the current use of the word 'lawn' is a bit skewed in one grassy direction. Indeed, what is a lawn? The rather bland definition found in most dictionaries is just not sufficiently comprehensive to give us a full and useful answer, a lawn clearly isn't just 'an area of short, regularly mown grass in the garden of a house or park' as the Oxford English dictionary would currently have us believe.

If I might be so bold as to brush away some stifling horticultural cobwebs (of which there are more than a few) and make a modest suggestion to the writers and compliers of dictionaries: if 'man-spreading' and 'hyperconnected' can make it into the Oxford English Dictionary as relevant words, then perhaps a revision of a well-known, well used, but evidently poorly defined word, is also timely and just as relevant.

THE VALUE OF LAWNS

Before we delve into the tapestry lawn format, it is worth mentioning that lawns in general have substantial value over and above their useful serviceability and superlative groundcover, but this value when viewed from an eco-friendly standpoint generally diminishes in line with the intensity of management applied; essentially the more you have to do to maintain your lawn and the closer it is to the glossy green 'ideal' or 'perfect' lawn, the less ecological value it tends to have due to the lack of plant diversity and the increasing amounts of energy and chemistry required to maintain it [11,12].

Nevertheless, lawns have what are known as useful ecosystem service features. These are the benefits they provide to humans simply by their existence and include acting as atmospheric nitrogen sinks [13], sequestering carbon [14], reducing rainfall run-off velocity and increasing run-off initiation time [15], acting as erosion control agents [16], via transpiration acting to moderate temperatures [17], offering both noise and glare abatement [18] and as a component of greenspace they may contribute to human well-being [19]. They also can influence property values and reflect societal norms [20,21].

Tapestry lawns have yet to receive the same degree of investigation into their potential benefits; however, it seems reasonable to postulate that none of these benefits is likely to be undone by the tapestry lawn format, and the research that has been undertaken to date strongly suggests that other additional benefits can also be accrued.

2 Tapestry Lawns
Form and Functioning

The familiar green, ground-hugging blanket of a typical low-mown grass lawn is achieved by many thousands of uniformly cut leaf blades of lawn-suitable grass species, and if managed appropriately can provide unmatched ground cover as lawns the world over demonstrate. Tapestry lawns (T-lawns) have the same fundamental purpose of providing low ground cover but additionally aim to increase the overall diversity of plants and concurrently the opportunities for the wildlife associated with them, as well as reintroduce flowers to an environment where they have been so recently discouraged. They are an ecologically informed, but nonetheless designed, plant community [22]. They don't happen by themselves.

Since several dozen format-suitable forb species are used rather than the usual restricted group of mowing-tolerant grasses, they produce a similar but distinctively different visual effect, one that is created by thousands of both cut and uncut stems and leaves, and at times includes flowers from plants that many of us will have probably grown up with but always regarded as weeds; it is a substantial change to the familiar grassy lawn format that can take a bit of getting used to (Figure 2.1).

With no clear precedent, it can be difficult to know what a tapestry lawn is supposed to look like. At the time of writing this, it is still highly unlikely that you will be able to look over your garden fence and make a comparison with your neighbour's T-lawn to see if things are looking about right. This is fresh twenty-first century, avant-garde ornamental horticulture. You are (if you'll forgive the pun) at the cutting edge, where ecological and plant sciences intertwine with horticultural art to the thrum of the mower and where there is still much to learn.

Even if you did peek over the fence and discover your neighbour had beaten you to it and had a T-lawn of their own, you would probably discover that your T-lawn looks a bit different. T-lawns simply can't be identically replicated for any length of time, and like many ornamental planting schemes, they may use the same ingredients but the outcome is always unique to the location.

Be mindful before you think to embark on creating a T-lawn of your own that they are not a lay and leave feature; like almost all created horticultural constructs, they will require ongoing attention. Mowing, weeding out the grasses, adding some new plants and taking out some bothersome ones is to be expected; it is primarily an ornamental garden-type feature after all. It has been suggested by another author that 'nature abhors a garden' [23], since inevitably natural processes are inexorably undoing all your hard work. So, fear not about having to aim for a finished or completed look or a 'perfect' T-lawn; just like any garden, your tapestry lawn will inevitably, and hopefully enjoyably, always be work in progress.

OUT WITH THE GRASSES

One thing that is conspicuous about highly managed traditional lawns and most of the modern grass lawn alternatives is the apparent desire to retain the look of a single plant-type monoculture. The simple construct of mown grasses that has become so ubiquitous throughout much of the modern human landscape has had a powerful influence on how lawns are perceived, but what would happen if the primary dominant competitors, the grasses, were removed from a typical NW European-type garden lawn?

FIGURE 2.1   A recently mown T-lawn shows how the lawn is made from a mixture of cut stems and leaves in a variety of shapes, colours and sizes giving a less mono-textured look than is found in well-managed grass-only lawns.

To many, we would be left with the lawn 'weeds'; the daisies, the buttercups, the clovers and the dandelions that had crept in uninvited, and that were certainly not present in the original pristine lawn. That there is space and opportunity in a grass lawn for more than just a limited selection of grasses is indicative; without a high degree of weed-killing management, traditional lawns can develop into a grass-dominated but nonetheless mixed plant community. The grasses are purposefully and exclusively put in place first by people, and then over time other opportunistic and format-tolerant 'weed' species follow.

You have probably encountered the simple garden adage that 'a weed is only a plant in the wrong place'. So, if we now turn the traditional format on its head and view the grasses as the weeds, and they (the grasses) are all then miraculously removed, we should have a collection of plants useful in a grass-free lawn. That is not entirely the case; not every forb found in a traditional grass lawn is necessarily useful to ornamental lawn horticulture, but what we certainly do have is a set of lawn plants that due to environmental filtering share a commonality in that they have all been able to withstand the rigours of survival in a traditional-type lawn.

Working with this refined group requires the usual approach to garden horticulture of having some understanding of individual plants and their needs, but is in itself insufficient; we also require an additional and informed approach to plant community management if we aren't planning on having a lawn jumble of mismatched or format unsuitable plants. Certain plants suit certain types of planting while other plants do not, and being aware of this in advance of creating a T-lawn is very helpful indeed.

With all the grasses removed there usually remains a rather diverse collection of forbs and usually some bryophytes (mosses). Depending on just how you view plants this scattering of little goodies might be appealing or not. It can certainly be challenging to modify a long-held viewpoint; moving from 'those damn buttercups' to 'those lovely buttercups' can take a while (if it happens at all), and not all these remaining botanical goodies are initially particularly desirable or aesthetically pleasing.

This was exemplified during early research on the development of T-lawns when an interested gardening group listened politely as the proposed format was outline to them and then they were presented with examples of the plants that were planned for inclusion. A pot each of common garden daisies, creeping buttercups, selfheal, red-and-white clovers and the like produced the comment 'It's a bunch of weeds', another comment was 'I have spent over 30 years pulling out buttercups from my garden and nothing you can say will make me want to put them back in'. There was much gentle chuckling at the seemingly crack-pot idea.

Considering the current trend for including 'natives' and 'wildflowers' in modern eco-horticulture this was a little disappointing, but not too surprising. Another set of pots that had been purposefully hidden under a table was subsequently revealed. It included red and pink semi-double daisies, a white-flowered creeping buttercup, white-flowered selfheal and red-leaved clovers with maroon flowers amongst other choice goodies. The members of the group crowded around and were rather interested. Words like 'lovely', 'pretty' and 'interesting' were used. They were the same plant species as originally shown but ornamental cultivars instead of what the audience already had in their garden lawns and borders, or saw frequently growing wild elsewhere. Viewpoints on what makes a plant worthy of a garden can be modified it seems, and that's a useful start.

Whatever your viewpoint, with the grasses removed entirely from a typical lawn it is the most successful of the uninvited plants and the spaces between them that are revealed. Dependent on your local conditions you might find yourself with a denuded lawn consisting mostly of vigorous forbs such as creeping buttercups, clovers, plantains and dandelions with some daisies and selfheal (Figure 2.2).

Fortunately, we are, or hope to be, lawnscape managers (lawn gardeners if you prefer) and are endowed with some sense of aesthetics, however individual that might be. We do not have keep all the plants that would remain, neither do we have to keep them as we might find them, which is generally as scattered individuals, small groups and larger colonies, although this distribution pattern and variable group size is of interest. We do however need to pay attention to the type and the behaviour of plants that would remain. The fact that they can spontaneously appear and creep into traditional lawns is a big clue as to the type of plants that are likely to succeed in a

FIGURE 2.2 Having out-competed a much-neglected grass lawn in a very wet and mostly shady position, a 'lawn' of creeping buttercups (*Ranunculus repens*) has taken its place. After being allowed to grow to their full height and then cut back, these regrowing buttercups reveal the space that can exist within a lawn. This space has the potential to be utilised by other plant species if the growth of the buttercups is purposefully managed.

lawn-type environment, and in selecting for the amount and type of plants that can do well in a tapestry lawn, it helps to have some understanding of plant biology, and to dip into some modern ecological thought for some management pointers.

Summary

- Horticultural lawns are created; they are not natural features.
- Tapestry lawns are managed plant communities.
- Tapestry lawns are not lay and leave features and must receive ongoing management if they are to succeed.
- Environmental filtering limits the types of plants that can survive in a lawn.
- Not all forbs that can be already found in traditional lawns are useful or desirable in tapestry lawns.
- Having some understanding of plant biology and ecological theory can be helpful in determining how to manage tapestry lawn plant communities.

PLANT TYPES

A study that looked at the composition of plant types in German lawns found that most of plants were of a type known as 'hemicryptophytes' (Figure 2.3). The word was coined by a very well-respected Danish ecologist by the name of Christen Christensen Raunkiær to describe plants with perennating buds at ground level [24]. 'Perennating' means to survive from season to season for an

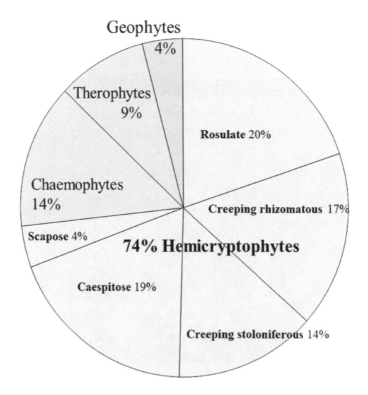

FIGURE 2.3 Life form spectrum of South Bavarian lawns. (From Müller, N., Lawns in German cities. A phytosociological comparison, in *Urban Ecology*, H. Sukopp, Editor, SPB Academic Publishing, The Hague, the Netherlands, pp. 209–222, 1990.)

indefinite number of years. Perennial grasses are examples of hemicryptophytes and so are daisies, dandelions, creeping buttercups and clovers.

Here we have a clue to our first important plant feature: possession of critical annual growth points (meristems) at the surface of the soil, which makes a great deal of sense when you think about it; the blades of a mower aren't meant to touch the ground and scalp the lawn, so fortunately perennating buds survive.

The other types of plants that may be found in traditional lawns are those that are lightly woody and have perennating buds just above the soil surface (low-growing chaemophytes) such as thymes (*Thymus* sp.); plants with short lifespans that survive unfavourable seasons such as winter in the form of seeds (therophytes), examples being scarlet pimpernel (*Anagallis arvensis*) and heartsease (*Viola tricolor*); and plants with underground storage organs (geophytes) such as the crocus (*Crocus* sp.).

The hemicryptophytes themselves can be further split into subcategories, which gives us further insight into other important plant features, namely growth form and how they are able to reproduce and spread in an environment where their flowers are frequently mown off and seed set cannot be guaranteed. Almost inevitably clonal reproduction and a propensity to produce adventitious roots where stems touch the ground is a particularly common although not unique feature.

There are the creeping soloniferous plants such as white clover, the stems of which creep across the soil surface and root at leaf nodes, and daisies which are known as being 'shortly stoloniferous' since its patches of stolons remain closely attached to the mother plant, only breaking off and becoming separate plants when it dies. Creeping rhizomatous plants such as yarrow (*Achillea millefolium*) have root-like stems with a shallow horizontal spread just below the soil surface, although their purpose is eventually to turn upwards and produce flowering stems and leaves rather than head down and take root.

Caespitose plants will grow in close bunches or produce tufts and are exemplified by the grasses, while rosulate plants have basal leaves arranged in the form of a rosette around the reproductive stem, such as hoary plantain (*Plantago media*), and scapose plants like dandelions (*Taraxacum* sp.) have a leafless floral stalk growing directly out of a strong root (Figure 2.4). Lastly, the geophytes are those plants with underground storage organs, such as tubers and bulbs, and include early-spring blooming lesser celandine (*Ficaria verna*) and the snowdrop (*Galanthus nivalis*). Raunkiær classified many other plant life forms, but it is these few types that hold the most useful and format relevant species.

Having largely identified the type of plant life forms suitable to a lawn-type environment it would be an understandable error to then generalise this selection and think of them as only small or low-growing plants. Anyone who has ever seen creeping buttercups grow unchecked will have noticed that the plant that is kept relatively small in a lawn by the repeated action of the mower can grow substantially larger in a border given the chance, easily to over 30 cm (1 ft tall), and yarrow left to its own devices can reach over 60 cm tall, and sometimes even taller still, with both of these species growing to these heights quickly in just a matter of weeks; another feature worth taking note of.

Plants that can survive the direct, repeated, and indiscriminate cutting by a lawn mower also tend to have a strong regenerative capacity; just think of how quickly the lawn daisy is back in flower just days after being mown. They are relatively fast growers and tend to allocate a lot of energy into producing their reproductive stems both floral and vegetative. Such plants will also often exhibit, to varying degrees, a survival strategy known as plasticity of form – the capacity to modify their growth pattern in response to major environmental influences. For example, a frequently mown yarrow plant will likely have substantially smaller and shorter floral stems compared to an unmown plant (Figure 2.5).

Plants that are unable to express this adaptive capacity to any useful degree eventually fall victim to the forces of environmental selection, and they tend to become locally extinct unless they have some other successful survival strategy; whereas plants that can usefully express this capacity will tend to survive and their offspring may carry this and other successful characteristics into subsequent generations. This is something of a simplification, but is largely how a plant species goes on

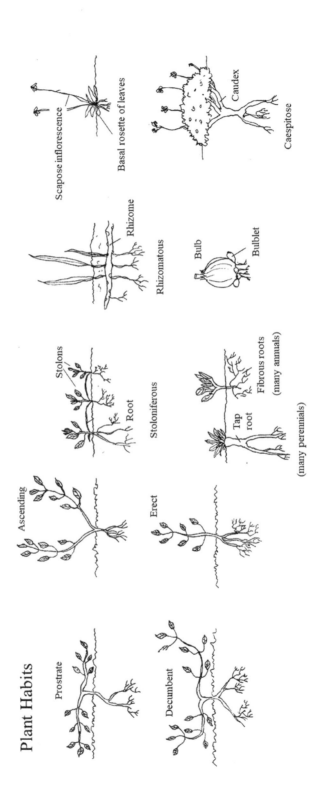

FIGURE 2.4 Plant habits.

FIGURE 2.5   More usually reaching 60 cm tall with colourful cultivars often found in the herbaceous border, here *Achillea millefolium* (yarrow) in a frequently mown and well-trodden utility lawn demonstrates plasticity of form by growing and producing flower heads almost as a prostrate plant. Unmown and untrodden, this plant is likely to regrow much taller and have substantially taller floral stems.

to develop phenotypes – forms of the same species that exhibit environmentally modified characteristics such as being prostrate or upright (Figure 2.6).

White clover is a good example of this with particularly tall (30 cm) and large-leaved forms known as 'ladino' varieties being used as animal forage in pastures and meadows, and smaller (10 cm) wild white or 'Dutch' varieties commonly found in heavily grazed grasslands and lawns (Figure 2.7).

Red clover (*Trifolium pratense*) shows similar variability that can be particularly noticeable in T-lawns (Figure 2.8), although unlike its cousin white clover, it is not considered to be a true clonal species and does not generally spread by stolons or rhizomes, rather it relies on seed for self-propagation. It is one of the species that is slow to spread in tapestry lawns.

FIGURE 2.6   Two phenotypes of *Trifolium pratense* (red clover) showing a tapestry lawn unsuitable upright habit on the left and a more suitable decumbent habit on the right.

FIGURE 2.7 Mature leaf size variation in different phenotypes of *Trifolium repens* (white clover) with a large-leaved 'ladino' variety on the left, wild or 'Dutch' clover in the centre and the microclover cultivar 'Pipolina' to the right.

FIGURE 2.8 A half-mown T-lawn showing an unsuitable red clover phenotype gaining height.

Plants that are directly cut by repeated mowing also inevitably undergo changes to their initial shape by the removal of both stems and leaves. This can also influence a plant's structural allocation of resources, with both repeatedly mown or grazed plants tending to show a greater allocation of resources below ground. Perennial lawn habitants frequently tend to have a proportionally greater root mass than plants that are not repeatedly defoliated. Again, if you think about it this makes a lot of sense; if your top growth keeps getting mown off, it's a good idea, if you can, to put more investments in your roots.

SUMMARY

- Useful tapestry lawn plants tend to have perennating growth buds just under, at or close to the soil surface.
- Most but not all useful perennial plant habits are creeping stoloniferous, creeping rhizomatous, caespitose or rosulate and include early and late flowering geophytes.
- Successful lawn plants show the capacity to quickly regenerate after mowing and may adapt their growth pattern in response to being repeatedly mown.
- Some species will have both T-lawn-suitable and T-lawn-unsuitable forms.

DISTURBANCE

Mowing affects all the plants in the vicinity of its application. Firstly, it disturbs the plant community, that is, we have a happily growing group of plants and then along comes a great big mower that chops off and removes everything above the set cut height. This chopping is not selective pruning in the manner of deadheading, or even selective munching in the way that animal grazing is; it is simple biomass destruction. Mowing cuts height-exposed plants continuously back to the same low height, a sort of repeated levelling of the playing field to the same benchmark each time. It effectually interrupts plant succession; the low lawn is prevented from steadily developing further into a taller meadow with its own set of meadow plant species. In terms of habitat it is rather like a multi-storey building being built and then repeatedly partially demolished and turned into a temporary bungalow before being rebuilt, which would be a disturbing event for any of the inhabitants, especially if it happened once a week or once a fortnight.

Even for plants in much less frequently mown tapestry lawns, height becomes a relatively short-lived advantage that cannot be maintained for long. Height might even be considered detrimental once the hum of the mower's blades can be heard, since it is the taller growing plants that will bear the brunt of the mower's action and are likely to lose the greatest number of leaves and stems in the process. But mowing has advantages if you happen to be a low-growing plant, since the amount of light that can be usefully intercepted post-mowing is determined by the number of healthy leaves that remain below the cut height, particularly their efficiency and the area that they cover, and prostrate plants that may have been previously overshadowed by their taller growing neighbours will have their leaves with greater access to light (Figure 2.9). Competition for light amongst plants in natural communities is thought to be a significant influence in reducing their diversity [25]; with that in mind, maintaining access to light to all inhabitants of a T-lawn makes a great deal of sense.

In a plant community, mowing not only disturbs and restricts succession but it is also quite a stressful experience for all affected plants, both the tall and short of stature; it can lead to what scientists call 'sub-optimal performance'. Unsurprisingly the plants that have just had their tops cut off go into the plant equivalent of shock, and the low-growing plants that have been stressed by having spent time in the increasing shade need to get used to being back in the light; neither are operating at optimum efficiency.

Post-mowing may appear to an observer to be a rather still period of transition with only the least-affected plants continuing to grow, but it only appears that way; there is plenty going on.

Defoliated plants that have been made smaller by the mower may now be out of balance and have more roots than they need to supply their reduced above-ground architecture, and roots may shrink and give up their stored carbohydrates to fuel the plant while it does its best to repair the damage. These taller plants will likely have also undergone some structural changes. With leaves, stems and buds removed, these plants may have to rely on older and less efficient leaves for a while, particularly later in the year when vegetative growth is retarded, and dormant buds may be aroused from slumber, especially if apical meristems have been removed. Apical meristems are the dominant

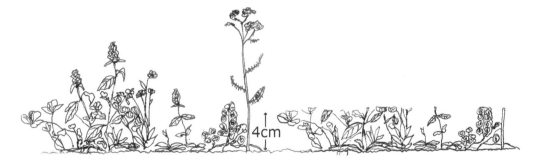

4cm

FIGURE 2.9 Taller plants overshadow lower-growing ones and eventually shade them out. Mowing at a height of 4 cm reduces the height of the taller plants, giving access to light to previously overshadowed plants.

growth points at the end of stems and shoots, and if removed, prompt side shoots develop. Mowing may initially result in a forb with a somewhat straggly or stubbly appearance but will often initiate regrowth that is both denser and initially less upright in form.

Previously shaded plants may have experienced etiolation (produced small unexpanded leaves and elongated shoots lacking in chlorophyll that can be a yellowish or whitish colour), and quite suddenly the freshly light-exposed plants may be receiving more light energy than they can successfully process into carbohydrate. In some circumstances where useful light has been excluded for some time and plants have produced shade adapted leaves, photo-oxidation may take place, a process that can result in the bleaching of some leaves and may prompt a newer set of sun-tolerant leaves to be produced.

Since this apparently still, but actually very busy, period is a cyclical occurrence and is going to happen to some degree after every mowing, it is a good idea to try for a balanced approach to mowing whereby plants are not allowed to become too tall or become subject to prolonged shade.

SUMMARY

- Disturbance is biomass destruction via mowing.
- Disturbance affects the entire lawn, even those plants not cut by the mower.
- Relatively tall and vigorous species tend to be most directly affected by mowing.
- After being disturbed, lawns take time to adjust.

MOWING

Mowing is essential if you wish to maintain a low aspect and have something that can be interpreted as a lawn rather than a low meadow or alpine-type lawn. It is also a behavioural signal that lets others know that it is a lawn you are maintaining and not some kind of low-growing garden bed or border since flower beds and borders are never mown (Figure 2.10).

FIGURE 2.10   Mowing is essential in the maintenance of lawns. It acts as an environmental filter and gives a clear signal that the space is managed and not neglected or running wild. The mown lawn on the right not only looks different, it is used differently, and its subsequent species composition and behaviour are also correspondingly different.

Mowing is a necessary management technique for tapestry lawns since it is the mechanism by which low creeping plants and relatively taller species can be caused to grow together in the same lawn community. It can be undertaken with any type of sharp-bladed mower, but importantly the mower must collect the arisings to prevent them from becoming a detrimental shade blanket if left on the lawn.

The frequency of how often a lawn will be mown will be largely determined by the plants themselves and by the sense of aesthetics and steely will of the blade-master. This is likely to be a somewhat irregular cycle if mowing is applied in response to increasing height several times a year. So far, mowing around three to five times a year tends to be a commonly applied figure in the UK, although this is dependent on the type of lawn being aimed for and can be weather dependent. This is nowhere near as frequent as grass-type lawns with their cycles of weekly or fortnightly haircuts, which in many suburban and managed lawns will be between 20 and 30+ times a year. Not only is this a labour and fuel-saving feature of tapestry lawns (and therefore also a reduction in CO_2 emissions in comparison to more frequently mown grass lawns), but the reduction in mowing frequency also allows for plants that would otherwise struggle to endure in a frequently mown grass lawn to have a much better chance of survival.

Mowing is required when the lawn begins to look like parts of it or all of it are starting to become an unbalanced, low hummocked meadow. It takes an aesthetic eye to make that final decision, but if certain plants are clearly beginning to overtop others to their detriment, plants are disappearing under others, or have begun to produce large hummocks of light-excluding patches, then it is time to bring out the mower. As a rough guide, if noticeably some of the lawn plants have managed to reach approximately 9 cm (3.5 in) in height and are acting to the detriment of their neighbours, then it is time to very seriously consider mowing the whole thing (Figures 2.11 and 2.12).

The temptation here is to be misguided by the current floral aesthetics of the lawn (Figure 2.13). If it is looking particularly floriferous or a certain species you have been waiting for has just come into bloom or is just about to, it can be very tempting to put off mowing, but fundamentally a tapestry lawn is a ground-covering lawn not a flower bed. The primary focus should be maintaining good ground cover rather than a floral display. Bear in mind that once you have started considering mowing it is probably because competition within the lawn has become visually apparent to the human

FIGURE 2.11   A boot in the lawn. If the plants are overtopping your shoes, then they are very probably in need of mowing. It can be tough taking a mower to a lawn spangled with flowers, but no mowing means no T-lawn in the long run.

FIGURE 2.12   Here *Veronica chamaedrys* (germander speedwell) has been just too pretty to mow and now covers a boot completely. It must be mown.

FIGURE 2.13   A freshly mown section of a T-lawn. The section to the right has reached a height whereby the lower growing and prostrate plants are being detrimentally shaded out. Having the steely resolve to mow off thousands of flowers is essential if the lawn is to maintain its species richness and remain a lawn.

eye and is already moving in certain species' favour. Some less vigorous or slower or lower-growing species is beginning to lose the competitive battle for space and light; any continued significant delay is likely to foster this situation. Not only will delay influence the composition of the lawn, usually to the detriment of the less competitive species, it may well leave you with particularly unsightly patches of rough-cut stems and bare soil where plants have been light deprived and died (Figure 2.14).

FIGURE 2.14 A freshly mown tapestry lawn showing a patch of cut chamomile that had been near to flowering and had been allowed to continue growing despite becoming a dense patch. After mowing, it is possible to see that there are no prostrate plants remaining within the patch. They been competitively excluded and grass-like cut stems and bare soil are now apparent.

The height at which to mow the lawn is informed by the basic requirement of all lawns in that it should function to maintain good low ground coverage; a bare or patchy lawn is rarely admired. For this to happen, plant survival is crucial. The set cut height should be one that gives the greatest number of plant species in the lawn the opportunity to survive rather than just act to reduce the height as required by a traditional grass lawn.

Mowing too low will tend to favour low-growing, prostrate and highly plastic species while mowing too high will tend to favour relatively taller ones, with competition subsequently likely to reduce overall species number in the lawn. The requirement is for a cut height that allows both types, both low and relatively high-growing species, to survive and coexist, one that is neither too high nor too short. Fortunately, experiments have been undertaken to determine a 'best fit' cut height, one that facilitates both the survival of low creeping species and taller species with the capacity to withstand defoliation by the mower and regenerate (Figure 2.15). The most effective cut in this regard has been determined to be around the 4 cm (1.5 in) height mark [26].

The removal of the arisings generated by mowing is also necessary to prevent the material forming a light-blocking layer on top of the lawn (Figure 2.16). This is both unsightly and detrimental to the continued development of the lawn.

T-lawn arisings can be treated in a similar manner to grass cuttings since they are mostly cut leaves and green stems (Figure 2.17). With sufficient volume, arisings can heat up very quickly in a compost bin. Within half an hour of cutting, three full hoppers worth of arisings can be hot to the touch and mixed with well-chopped straw can make a very fine garden compost.

SUMMARY

- Mowing approximately three to five times a year is generally useful in UK conditions.
- Any type of mower will do, but it must be able to collect the cuttings.
- Mowing is undertaken for aesthetic and ecological reasons and to maintain the low height and ground cover required of a lawn.

FIGURE 2.15   Arisings left by the mower on the lawn look unsightly, can stick together and block light to the covered plants.

FIGURE 2.16   A close-up of a tapestry lawn that has been freshly mown to 4 cm and the arisings removed by the mower. The flowers have been cut away and close-up there are cut leaves and stems to be seen; however, this cut height facilitates the survival of most lawn species and there is very little bare soil.

- Mowing is part of the repeated cycle of disturbance necessary to modulate competition within the lawn.
- Mowing to around 4 cm in height facilitates the coexistence of the greatest number of both low growing and relatively taller species.
- Arisings from mowing must be collected and removed to maintain a high light environment and potentially contributing to soil nutrition.
- T-lawns require no additional fertiliser.
- A T-lawn is a lawn; it is not a flower bed. It must be mown no matter how pretty it is looking.

FIGURE 2.17   Fresh arisings from a mown T-lawn. Mixed with high carbon sources such as well-chopped straw, the freshly shredded T-lawn arisings can be used to make a very fine garden compost.

COMPETITION

You will have probably noticed that all things being equal that smallish low-growing plants such as sun-loving thymes and the like tend to have many small leaves while larger growing plants such as buttercups tend to have significantly fewer but relatively larger leaves. Both are adaptations that in part attempt to optimise the amount of light that reaches the leaves when taking in to account the habit of the species and the environmental conditions they do well in, particularly the availability and efficient use of water (Figure 2.18).

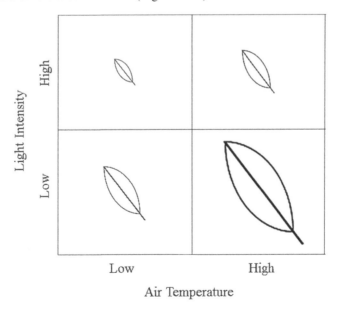

FIGURE 2.18   Based on water-use-efficiency, leaf size can frequently be related to light intensity and temperature. (From Parkhurst, D.F. and O. Loucks, *J. Ecol.*, 60, 505–537, 1972.)

Generally speaking, the larger the leaf, the greater its water requirements; shade and light intensity can influence this requirement, as can temperature, by speeding up or slowing down a plant's metabolism, and very low frosty temperatures tend to do more damage to larger leaves than to smaller ones. Large leaves can also be more photosynthetically productive and generate more carbohydrates than smaller leaves, and perhaps unsurprisingly larger-leaved lawn hemicryptophytes when supplied with sufficient water tend to be more vigorous than their smaller-leaved counterparts. They will however be amongst the first to show signs of stress if water becomes limited for any length of time.

With less leaf area for photosynthesis to take place, the smaller-leaved clonal lawn species and varieties tend to be less overtly vigorous; they are frequently low growing with numerous small leaves, and often spread by slowly creeping and generally consolidating their position in patches. The lawn daisy (*B. perennis*) is a classic British native example and creeping mazus (*Mazus reptans*) a relevant non-native example. The ecologist Lesley Lovett Doust named this type of creeping behaviour as 'phalanx'. Essentially the patch grows steadily outward in the manner of soldiers locked in close formation, as compared to the larger-leaved clonal species, which tend to have a different behaviour known as 'guerrilla' whereby stolons and rhizomes can stretch out and effectively jump over or bury under other plants and then assert themselves by growing faster and taller than their competitors; the creeping buttercup (*R. repens*) is a good example of a guerrilla-type plant [27].

It would therefore seem logical if you were to grow unmown daisies and buttercups together that without intervention the guerrilla tactics of the buttercup would likely prove the most competitively successful tactic. It could infiltrate the daisy patch, hijack its territory and shade it out by outgrowing it, unless the mower were introduced to repeatedly stall the invasion and regularly bring the buttercup down to size. This is relevant when planting a tapestry lawn since all plant species compete with each other for the limited local resources of space, light, nutrients and water, and as lawnscape managers we must inevitably wonder how much of any one species is appropriate to use in the lawn community mix.

The question arises, how many buttercup stolons would it take to completely overcome and monopolise a daisy patch? Inevitably the answer is largely dependent on the size of the daisy patch. A very large patch of daisies would withstand many more buttercup stolons. This competitive scenario gives us a useful clue in how to think about deciding on the proportions of different plants to use in our lawns. Would it make sense to have lots of buttercup plants? Probably not. Would it make sense to have lots of plants with attributes similar to buttercups? Again, probably not. Plants like daisies however we can probably use lots of.

Why all this attention to stress and competition you might ask? It's not something you find much of in other gardening books. Another dip into ecological theory may be useful here. Although it continues to be a contentious topic amongst ecologists, there is the suggestion that all things being equal, that we might expect the highest levels of biodiversity to occur when there is a balance between stress and disturbance on the one hand and competition for light and space on the other [28].

Unlike traditional gardening methodology where the aim is usually to keep individual plants as stress and disturbance-free as possible so they individually thrive and in their season produce a good show, in the case of a successful tapestry lawn where we are working with a mixed community rather than choice individuals, we need to ensure we manage the community, rather than just the individual plants, and include both disturbance and stress in a sort of 'treat them all mean to keep them all keen' style of management approach. It is an analogue of what might occur to munched-upon plants in the wild, and it is worth remembering this when the plants behave more like wild-type plants than the cultivars that they may be; a bit of that phenotypic plasticity mentioned earlier will probably be in play. The ornamental daisy in a pot or bedding display will look rather different to the same type daisy growing in a mown and highly competitive tapestry lawn community.

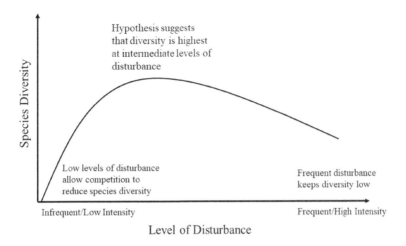

FIGURE 2.19 The intermediate disturbance hypothesis suggests that greatest diversity is likely to occur at intermediate levels of disturbance. (From Connell, J.H., *Science*, 199, 1302–1310, 1978; Grime, J.P., *Nature*, 242, 344–347, 1973.)

The influence of disturbance on biodiversity is frequently shown in a graph and known as the Intermediate Disturbance Hypothesis, a theory proposed by the American ecologist Joseph H. Connell after observing the behaviour of barnacles on the coast of Scotland (Figure 2.19). The hypothesis suggests that a bit of disturbance encourages diversity, but that in environments where disturbance is very frequent and intense or rather infrequent and weak, very few species will be able to survive [29].

An example of intense garden disturbance would be digging over a patch of soil once a day for a few months and then seeing how many vascular plant species manage to live and reproduce in that patch, which would probably be none; even short-lived ephemeral weeds wouldn't manage it. The same applies at the other end of the spectrum where disturbance is rare or mild and the most competitively successful species are likely to come to dominate all others. This a derelict garden-type scenario; if the garden isn't regularly disturbed by the actions of the gardener doing a spot of gardening, then the highly competitive weeds are likely to take over, followed by competing shrubs and fast-growing trees, and then finally long-lived trees shading out all of them, with the most successful long-term competitive tree species becoming completely dominant.

This hypothesis suggests that diversity might be maintained in tapestry lawns by the relatively infrequent but locally intense application of the mower. Although it is important here to remember that tapestry lawns are not the natural and semi-natural communities that the theory attempts to give some understanding to, they are artificially created horticultural features and likely to be directly modified over time by 'lawn gardening' as plants are purposefully added and taken away. It is also important to remember that tapestry lawns are not an attempt to re-create or restore a natural environment, but to expand upon an existing man-made landscape feature by constructing a plant community that meets the prevailing requirements of maintaining ground cover, offering aesthetic interest and providing environmental enrichment.

Ecological theory has even more to offer here to help us understand how things might function in tapestry lawns. If we think of the creeping buttercup, we might say its strategy for success is to be a short-lived (most buttercup stolons rarely live beyond a year or two), but a highly productive guerrilla-type plant that out-competes other plants by reproducing quickly, growing fast and becoming taller than its competitors, and it can recover from the occasional disturbance of mowing. Other plants can and do have different strategies.

Daisies tend to be found as tight, slowly creeping phalanx patches in very low or cut vegetation such as lawns, they rely on the disturbance of mowing or grazing to provide the well-lit conditions that suit them, they don't take well to being stressed, that is, shaded out, but they do go largely unscathed during mowing. The disturbance of mowing that affects the surrounding community literally goes over their meristems and is actually crucial for their survival. This demonstrates that plants have different strategies in their approach to survival in environments similar to that of the lawn, and we may ask, is there a best strategy for survival in the artificially created and disturbed community of a tapestry lawn?

The answer is essentially 'no'. In terms of plant evolution, lawns have not been around for even half a blink of an eye, and survival strategies are essentially 'borrowed' from analogous environments. However, certain plant strategies have been proposed for natural communities by the prominent British ecologist J. Philip Grime in the form of the Universal Adaptive Strategy Theory [30,31], and it is the variety of survival strategies that can be observed in tapestry lawns that are of interest to us.

In brief, the theory suggests that surprise, surprise, plants have evolved survival strategies in response to both stress and disturbance. Here we should clearly define stress to mean anything that can restrict plant growth, such as shortages of light, nutrients and water or extremes of temperature, and disturbance means anything that destroys plant material, such as mowing, grazing, fire and football matches. Plant communities can experience both stress and disturbance to any degree and in any combination, and while low or moderate combinations of stress and disturbance can support vegetation, extreme combinations cannot (Figure 2.20).

This can be further refined to indicate the kind of survival strategies that different plants use, with them being thought of as stress-tolerating strategies, ruderal strategies, or competitive strategies (Figure 2.21). Plants may exhibit a single strategy type or combinations of types (Figure 2.22).

As examples using this system, creeping buttercups show a CR-type strategy and daisies show R-CSR-type strategy. A look at CSR classifications for other T-lawn plants shows that plants with all kinds of survival strategies can be used in the artificial world of T-lawns, with mixed survival strategies being more common than single ones.

If this all seems just a little bit overly complicated just for choosing plants for a lawn it is worth knowing that most ecologists would agree with you, and your author would heartily agree with them. However, it is outlined here since plant community management will be new to most gardeners, and the theory does reinforce the importance of stress and disturbance on plant survival and behaviour, and by extension their importance on plant communities, natural or otherwise. Reassuringly, there is a list of tried and tested plants in the final section from which to choose when creating your own lawn community.

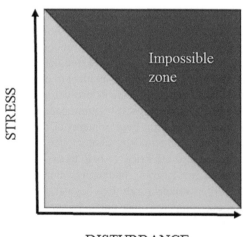

FIGURE 2.20   Plant survival under the influence of stress and disturbance is only possible within the green zone. Existence within the extreme zone shown in red is not biologically possible.

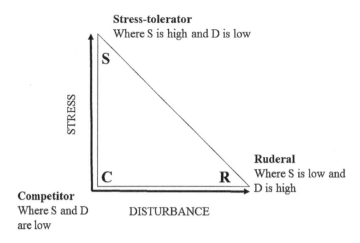

FIGURE 2.21   Major plant survival strategies: Competition, stress-tolerance and ruderalism. (Grime, J.P. et al., *Comparative Plant Ecology: A Functional Approach to Common British Species*, Springer, 2014.)

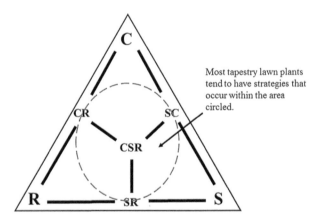

FIGURE 2.22   The CSR triangle showing some of the intermediate strategies.

FACILITATION

There is another side to the competitive coin worth mentioning and that is facilitation. This is when one plant can influence the survival of another in a positive manner, rather than the potentially negative outcome of competition.

After noting how competitive shading can be detrimental to low-growing and prostrate plants in a T-lawn, it may be a bit confusing to learn that in some circumstances, usually related to the time of year, that a bit of shading is a generally benign event for some of them. It is only some plants that benefit.

Species adapted to strong light such as thymes do best in the full sun at all times, but for those plants that can happily survive in varying degrees of shade or tend to dehydrate with dry days and full sun exposure, these may benefit from the shade and trapped moisture facilitated by the overtopping leaves of their competitors (Figures 2.23 and 2.24).This type of shading is most benign during high and late summer when light levels and grass-level temperatures are at their greatest.

On the topic of grass-level temperature, it is worth knowing that air temperature in general is measured in several ways, and one measure that is not often mentioned is particularly relevant to lawns and gardens.

What is meteorologically referred to as the 'grass minimum temperature' is a record of temperature in open ground on short turf, with the bulb of the thermometer just in contact with the tips

FIGURE 2.23 *Ajuga reptans* 'Braunherz', a plant of woodlands is here being partially shaded under a mid-summer foliage canopy created by white clover.

FIGURE 2.24 *Ajuga reptans* is generally a woodland and woodland edge plant of dappled shade. In high summer, the mixed shade provided by the clover can act as protection against the sun's rays and can also help maintain a narrow ground-level humidity layer that further ameliorates the environmental stress. Although competing with the *Ajuga*, the clover is at the same time also potentially facilitating its continued existence during what can be a stressful period, particularly in an open and well-lit lawn environment.

of the blades of grass. It is also described as the temperature at 5 cm above ground, and is therefore just above the T-lawn recommended cut height. This temperature often varies substantially from 'air temperature', which is measured at 2 m above ground and is usually reported in weather forecasts. The grass temperature can be much higher than the air temperature on warm sunny days and much, much lower (e.g. ground frost) during clear weather under a clear winter sky. The point here is that lawns can become substantially warmer or substantially cooler than the air that is above them and therefore be subject to much greater temperature variations than their surroundings. It can be a tough temperature environment for many plants.

Plants regulate temperature via respiration, losing water in a manner similar to, but distinct from, sweating. However, at high temperatures the physiological process within a plant can simply cease to function effectively. Although a bit of stress in a tapestry lawn is generally a good thing, a high grass temperature can cause a plant to 'shut down' in an attempt to conserve water, and it can become heat damaged as a result.

In the dense but effectively shallow planting of a T-lawn a bit of shading can be beneficial to those plant species that are light and temperature sensitive, especially if conditions tend to be dry. Plants that benefit from a bit of summer shade tend to be forbs with early spring-flowering periods that may have their wild origins in woodland-type conditions, plants such as primroses and *Ajuga*. It is possible to benignly overlook, but not completely ignore, the extended period of shading that may occur between mowing events if it is only this type of plant that is being affected. This bit of purposeful, but temporary, disregard for competitive height can also allow for later and taller-flowering species such as *Pilosella aurantiaca* (fox & cubs) to achieve the height necessary for their late summer floral display if the other T-lawn constituents allow for it. If, however, you should see your thyme struggling, it will be time to mow.

SUMMARY

- Use greater numbers of creeping plants with small leaves.
- Aim for a general balance between stresses such as shading and disturbance such as removing the shade.
- Don't expect plants in tapestry lawns to look or behave like the same plants grown in borders or containers.
- Delaying mowing in hot, dry weather is allowed.

DIVERSITY

Humans may to have an innate capacity to perceive biodiversity positively [32,33], and having a variety of plants is not only visually appealing, it is also more likely to produce an aesthetically useful display over a longer period if the plants are well selected. It also produces a varied-looking plant community; it would be a rather lacklustre lawn if there were only daisies and buttercups. If we consider the buttercup and the daisy example, it is apparent that not all plants are useful in equal amounts, so the variety of plants to use and how much of any one plant species to be used at inception are relevant questions, and we might even ask why go to the bother of having buttercups at all if they can be thugs?

To address the subject of buttercups, it is necessary to use a relevant if rather imprecise, horticultural term. Buttercups for all their apparent drawbacks are 'doers'; they are robust and can look attractive when in bloom, not to mention that in the British Isles they are a valuable botanical resource to over 30 different species of insects, including many pollinators [34].

'Doers' in the tapestry lawn sense means they:

- Contribute to the ground cover essential for a lawn.
- Remain in leaf during an average British winter.
- Provide floral resources and display.
- Can withstand the disturbance of regular defoliation.
- Act to competitively stress other plants.
- Contribute to the total number of species and the overall plant diversity within the lawn.

FIGURE 2.25 Doing more. Two types of ornamental creeping buttercup (*R. repens*), the pale star-flowered cultivar 'Gloria Spale' and in smaller numbers the double-flowered 'Flore Pleno' with a mix of other T-lawn species.

'Doers' in tapestry lawns have to 'do' far more than just flower reliably and prolifically like the traditional doers found in a typical garden border, window box or hanging basket (Figure 2.25).

As there are no plants that have evolved to inhabit ornamental lawns, we must therefore cannily select each species from a pool of hemicryptophytes that have evolved to suit other more natural and, in some way, related habitats. The amount of each species we use for our new designed community will necessarily be informed by a plant's habit, its characteristic traits and survival strategy, and also by what kind of environmental and competitive conditions it tolerates and those it can do well in.

Since each plant species has evolved to survive in particular conditions, it will fit what is known as an 'ecological niche'. An ecological niche is something of a traditional idea in ecology that has in recent years been challenged and shown to be of limited value; however, the concept can be broadly useful and a niche is in essence the role and the position a species has in its environment that fosters its survival, that is, how it meets its needs for nutrition and protection, its survival strategy and how it reproduces. It may be an old idea but it can be helpful.

The plants best used in tapestry lawns all have the capacity to inhabit ecological niche environments that can be at least partially provided by, imitated, or simulated in, the occasionally mown tapestry lawn, and considering that they can all survive in lawn-ish-type environments their niches are often likely to overlap, even if they wouldn't meet naturally; for example, plants from New Zealand wouldn't naturally meet plants from Europe. The more overlaps there are between species, the greater the competitive jostle for space and resources, and fortuitously for a lawn this means that it is less likely that there will be much bare ground, since many species will be competing for it, which from our point of view is a useful and desirable situation. All the jostling for space and resources may also go some way in helping to maintain the overall stability of the lawn community and help in reducing the likelihood of any one species becoming overly dominant. To understand this better it is once again helpful to dip into ecological theory.

If we were to selectively look at just light and moisture as two influencing factors and imagine a hypothetical plant that is something akin to creeping thyme (*Thymus praecox*), and we broadly classify it as a plant that was suitable for a relatively dryish sunny spot, it would have an ecological

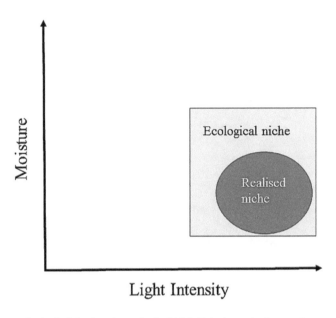

FIGURE 2.26   The ecological niche for a hypothetical high light intensity/low moisture type species and the realised niche it occupies when in competition with plant species that can inhabit the same niche.

niche that would encompass these conditions (Figure 2.26). It likely has the potential to survive in a sunny and dryish-type lawn. However, if it is in competition with other plant species with shared tolerances and similar strategies, it is unlikely to be able to fill the entirety of its ecological niche; the other plants in their fight for space and resources may competitively exclude it and, in some locations, do better. It might come to only inhabit what is known as a 'realised niche', which is the sunny dry space it can inhabit when in competition with these other species, its ecological niche being a bit wider when it does not have any competition. The same would apply if we were to add another hypothetical but more moisture-tolerant species with similar light requirements (Figure 2.27).

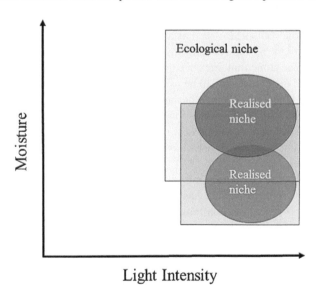

FIGURE 2.27   Ecological and realised niches for two hypothetical high light intensity species. Both species' ecological niches overlap; they could occur within a similar type of habitat, but with their different moisture tolerances and competition between themselves and their neighbours, their actual realised niches would likely be different and variable.

FIGURE 2.28 In the shared T-lawn, habitat species with overlapping ecological niches can often be found to have self-organised and be growing together as the lawn develops over time. Having an understanding of the immediate environment that the lawn provides can be useful in selecting initial species that will likely prove successful. Here in a relatively well-drained and sunny spot *Acaena microphylla*, *Ajuga genevensis* × *reptans*, *Argentina anserina*, *Leptinella potentillina*, *Pilosella officinarum*, *Thymus praecox* and *Veronica chamaedrys* have intermingled in a three-year-old T-lawn in the South of England.

In effect, there are likely to be several different species that can occupy a particular habitat. In T-lawns this may become apparent over time as the lawn develops, with species that share similar ecological niches often tending to occur together in a veritable mêlée (Figure 2.28).

Small changes in the immediate environment can become influential in plant survival; a slight change in the level of the soil surface such as a dip may give added protection from the mower for one species, while it might be the cause of a little extra shade or improved moisture availability for another. A subsurface stone may improve drainage for one species or restrict root development in another. These small differences are to be welcomed since they contribute to improving the variability in what has the potential to be a rather monotonous terrain.

Environmental complexity such as variations in seasonality, light quantity and quality, temperature, moisture availability and retention, soil texture, organic matter content, microbial activity, compaction and aeration, drainage, nutrient availability and frequency of footfall can all make a difference (Figure 2.29).

Additionally, within the plant community itself the competition between plants both individually and in combination with each other has a profound influence; ecologists tend to call this type of internal community competition 'community matrix theory'. Community matrix theory is based around the idea that competition in communities with many species can be separated into two processes: the effect that a species has on all others, and the response it has to all others, and that there is a hierarchy of competition with some species effectively stronger than others and known as asymmetric competition (Figure 2.30). It also suggests that under certain circumstances, richly diverse communities can reach a type of stability whereby if the community is disturbed it can in time revert to something similar to its pre-disturbance condition. This is known as the diversity-stability hypothesis; essentially the theory suggests that the more species there are in a naturally derived community the more robust it tends to be.

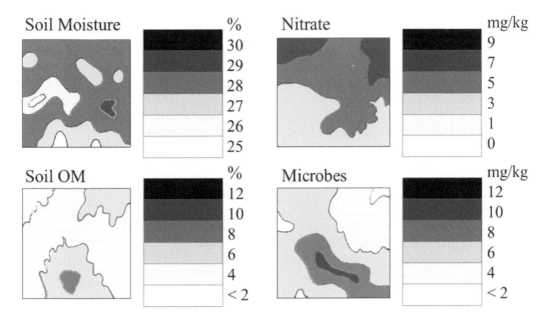

FIGURE 2.29   An example of the type of variability in just four soil parameters that can be found in 1 m² of grassland.

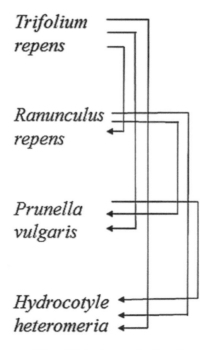

FIGURE 2.30   A hierarchy of competitive abilities for species found in a traditional lawn. An arrow leaving a higher species and terminating at a lower species indicates that the higher species suppresses the lower species significantly more strongly than the lower species suppresses the higher one. (Adapted from Roxburgh, S.H. and Wilson, J.B., *Oikos*, 88, 395–408, 2000.)

However, once again we are delving into the ever-contentious behaviour of natural and semi-natural communities rather than those of attentively constructed tapestry lawns, but we might make the cognitive leap that the more species-rich a tapestry lawn is the more gradual that any degradation is likely to be. There is some strong evidence that suggests that diversity-stability relationships are strongly context dependent, but the idea can be useful.

Our context is the designed community of a tapestry lawn rather than a natural community, and we are not trying to restore, replicate or conserve a natural habitat but rather enhance an entirely manmade one by making it more biologically diverse. However, we can utilise this hypothesis by using as many suitable species in a tapestry lawn as we can, in the hopeful expectation that with all those format-suitable species jostling for their different niches, combined with continuous annual variations in the non-biological conditions such a temperature, rainfall, time of year and the competition-levelling action of the mower, that at any one time some conditions will be favourable to some species and less favourable to others and that this will be broadly cyclical and ongoing. One year we might find that the buttercups do particularly well and the next year it is the daisies.

To date, limited experiments with different numbers of format suitable species in tapestry lawns tend to support the diversity-stability hypothesis and suggest that in the short-term as total species number increases, so the rate of proportional change within the lawn can decrease (Figure 2.31). Essentially the more crowded your lawn is with different species, the more stable a community it has the potential to be, although not all of the species may endure. The experiments were inevitably limited to the small number of selected species they used and are more of a window into what can happen rather than what will happen, but they do suggest a horticulturally useful trend if you plan on having your species-diverse lawn around for a while.

Using a wide variety of different plants also has an influence on the overall ground cover with lawns with more species showing more complete coverage (Figure 2.32).

It is worth remembering that in temperate regions of the world the change of season will also impact coverage. Some of the species that are suitable for tapestry lawns such as silverweed

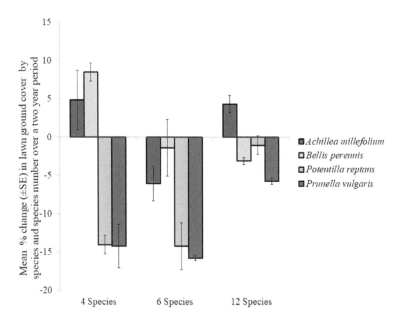

FIGURE 2.31   As the species number within a tapestry lawn goes up, the amount of overall variation may be reduced. This kind of behaviour is also likely to be influenced by the species' choice, the overall mix and proportions and the environmental conditions. (From Smith, L.S. and M.D. Fellowes, *Landsc. Ecol. Eng.*, 11, 249–257, 2015.)

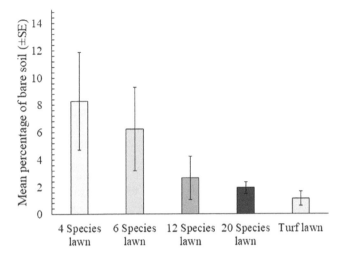

FIGURE 2.32 Over time, the amount of noticeably bare soil will vary in a tapestry lawn. The variation and amount of bare soil generally decreasing with increasing species number. Experimental results over 2 years (shown) suggest that lawns of 12 or more forb species have the potential to provide coverage similar to that found in a traditional grass lawn. (From Smith, L.S. and M.D. Fellowes, *Landsc. Ecol. Eng.*, 11, 249–257, 2015.)

(*Argentina anserina*) are deciduous; they seasonally lose their leaves and provide autumnal colouring as they fade but may leave noticeable space behind them if the T-lawn sward is not particularly dense. Some like white clover are partially evergreen and only tend to lose their leaves when temperatures fall more than a few degrees below zero, and some like yarrow (*A. millefolium*) remain evergreen, while others such as creeping mazus (*M. reptans*) retain their leaves but may experience a leaf colour change, often turning shades of reddish-brown as protective pigments come into play. Seasonally this can produce unusual multi-tonal lawns (Figure 2.33).

FIGURE 2.33 A multi-tonal lawn late in the year when most peak floral periods have passed. The brownish and reddish leaf colour is mostly from cultivars that already have this colouration, although many of the T-lawn plants will respond to falling temperatures by darkening.

If we were to lose one species from a mix of thirty lawn plants, then the community is unlikely to be overly affected unless the plant happens to be a notable competitor (although we are already manipulating this via selecting from within a suitable range of species and doing some mowing). Loss of a species within a tapestry lawn is an indication that the competitive mix it is encountering and/or the environmental conditions it has experienced are not suitable to its ongoing survival. It is of course quite possible to do a spot of lawn gardening and replant the lost species if you should choose to give it another go or maintain its presence no matter what, but it is likely that unless there are changes to the immediate competitive mix (i.e., try putting it in another spot with different plants), or the environmental conditions (i.e., try a more appropriate situation), that the same disappearance will likely loom large again.

Most lawn extinctions tend to occur when a species or cultivar has been rather optimistically included in the mix, even though they are not suited to the prevailing local conditions, such as using a thyme (*Thymus* sp.) in a spot outside its ecological niche, perhaps in a damp and/or shady part of the lawn. It is worth giving some attention to where in your lawn you initially place your plant ingredients; damp spots, dry spots, trees, light and shade are worthwhile noting in advance (Figure 2.34), and plant choices and subsequent planting undertaken accordingly.

Extinctions can also occur by planting or fostering too many vigorous plants such as creeping buttercup, and making the lawn too competitive, or initially planting too small an area of slow-to-spread species so that they are unable to properly establish when the lawn is first planted. Occasionally some plants will disappear after encountering a pathogen such as the Australasian rust fungus *Puccinia lagenophorae* now found on some ornamental daisies across Europe, although diverse communities like tapestry lawns are thought to be less prone to severe pathogen and pest damage.

FIGURE 2.34 A self-organised community of plants under a field maple (*Acer campestre*). Some of the original species planted have moved on or become locally extinct leaving a grouping of plants able to survive the immediate conditions. Here *Geranium thunbergii*, *Ajuga reptans*, *Leptinella squalida*, *Glechoma hederacea*, *Viola hederacea*, *Veronica chamaedrys* and *Trifolium repens* 'Pentaphyllum' blend to give good ground cover in a seasonally shady and occasionally dry location.

HEIGHT CONVERGENCE

Some intriguing behaviour has been observed in plant communities that suggests that some communities exhibit what is known as height convergence [35,36]. This is where plants in crowded situations (such as tapestry lawns perhaps?) are observed to generally regulate the height to which they grow to be similar to that of their neighbours, even when there may be advantages to growing taller (Figure 2.35). Ever noticed how in profile that woodlands with many different types of trees seem to reach a roughly similar height?

There are a variety of potential explanations to this behaviour involving such exciting topics as light use efficiency, photomorphogenic signalling and biomass allocation, but what is relevant to tapestry lawns is that for a useful period the lawn community can appear to behave itself very nicely, with most of the plants apparently content to remain roughly the same height as their neighbours given that they have similar conditions.

Inevitably there will be moments when this apparent bonhomie fails, and a plant will make a break for it and reach for the sky; usually although not always, this is when it wants to get noticed and be pollinated. Tapestry lawns require this wayward behaviour of leaf and flower to induce mowing, and you now know how important that is. It also has another plus point if you are prepared to allow your lawn to have a bit of height variation and to express tall floral stems, particularly in mid and late summer when yarrow, mouse-eared hawkweed and fox & cubs can come into bloom. The floral stems are generally narrow and few and certainly can be tolerated, just so long as the rest of the lawn is still behaving.

SUMMARY

- Competition between plants can facilitate good ground coverage.
- There is a hierarchy of competition with some plants being more robustly competitive than others.
- Most mobile plants will tend to come to inhabit the most equitable location in the lawn according to their niche requirements and the prevailing environmental conditions.

FIGURE 2.35 Mowing is the great leveller in any lawn. For a while after mowing, the diverse species in the lawn appear to maintain a similar overall height as they regrow, with just a few emergent flowers wanting to be noticed poking above the induced canopy. This intriguing behaviour usefully maintains the expected low-cut appearance of a lawn.

- The potential stability of the lawn community may increase with the number of different species within it.
- Not all plant species will survive the ongoing competitive battle if the prevailing or seasonal conditions don't suit their niche requirements.
- If plants are being consistently overwhelmed or a few species are becoming dominant, then a spot of lawn gardening is probably required.
- The lawn may keep to an overall similar height for a while before errant plants begin to pop up.

PROPORTION OF PLANTS TO USE

Observation of natural plant communities indicates that many are dominated by a few successful species and that most other species tend to occur in smaller numbers. The species with the greatest number of individuals (abundance) is said to have the highest rank, the second most abundant has the second rank and so on. This is usually illustrated as a species rank abundance curve (Figure 2.36).

Many of the new approaches to aesthetic horticultural communities are guided by ecological principals [22,37–40], and as you may have noticed, in many ways tapestry lawns are too. However, in a tapestry lawn we are not seeking to create an analogue of a natural community or how it functions. It is an artificially derived and artificially maintained community with a specific horticultural purpose, and to give it the best chance of fulfilling this purpose, we need to bend the rules that are thought to be at work in natural communities.

The reasoning here is simple: if we had lots of the most dominant species in our lawn, that is, the biggest and fastest-growing plants, it would not be particularly diverse and may not be particularly attractive. It was earlier postulated how unmown creeping buttercups and species with similar characteristics might take the role of a high-ranking species and competitively eliminate others in a hypothetical lawn of daisies, so it becomes apparent that we should manage the type and amount of any one plant that we use. Remember how we would need a very large lawn with lots daisies to have any chance of competing with a few creeping buttercups. We must forego the structure found in many natural communities and with benign horticultural purpose we must invert the rank abundance curve (Figure 2.37).

The question then arises as to which T-lawn-suitable plant species would fall into which rank abundance categories. This is a question that inevitably has no absolute answers since we are working with populations of living organisms, in variable conditions, that often include cultivars particularly selected by humans for their unusual characteristics and horticultural merit, so all we can really do is make a broad comment on how most of the population behaves, most of the time; by necessity we must broadly generalise.

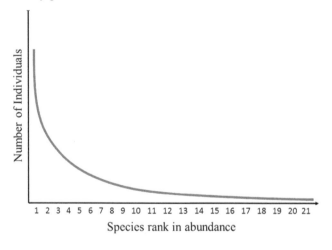

FIGURE 2.36 A generalised species abundance curve. Many natural communities are dominated by only a few species in large numbers while most other species are represented by significantly fewer individuals.

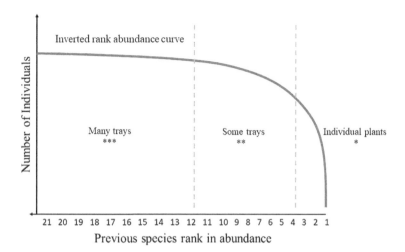

FIGURE 2.37 Inverted rank abundance curve. Dominant species are reduced in number and become fewest while greater numbers of low-ranking species are initially used. For planning and planting purposes, the use of individual plants*, some trays** and many trays*** is indicated (See section on plant tiles). By this method, we introduce more of what would be classed as low-ranking species in substantially greater numbers.

Fortunately, as outlined earlier there are some characteristics, survival strategies and features that can give useful pointers, the general leaf size for example, with smaller-leaved forbs tending to be less vigorous than larger ones. Excepting some cultivars, the average height that a forb might reach tends to be greater in vigorous species and forms, and the concentration of chlorophyll is another frequently relevant feature, with darker-green-leaved plants tending to be more productive than pale or variegated leaf forms.

Here the botanical ecologists will probably be cringing in horror at what might be considered at best an oversimplification, and they might be disposed to note that the age of the plant and its individual leaves, the leaves' positions on the plant, its degree of exposure to direct sunlight, its environmental location, its level of nutrition and water availability, its phenotype, its ecotype, the plant's bacterial and mycological associations and a surprisingly long list of other factors might also be very relevant. They wouldn't be wrong, but mercifully we are dealing with choosing the amounts of different plant species for a species-rich ornamental lawn, not a treatise on forb biology, and any-way, we are free to subsequently tinker with our lawn if it looks like things aren't turning out quite how we would like.

It would be improper to suggest that there is a 'correct' amount of any of the species to use since each T-lawn will continuously change and develop as the plants self-organise in response to the local environment, disturbance and competition. All we are trying to do is give the T-lawn a good start and prevent it from being overrun by vigorous species. As the inverted rank abundance curve suggests, at its simplest it is probably best to use lots of less-vigorous plants and substantially fewer numbers of more vigorous ones if we don't want to end up tinkering too much. Experiments suggest that with the exception of some cultivars, that notably vigorous T-lawn-suitable plants with a high rank abundance of 1–4 should occupy no more than around the 5%–10% mark of a tapestry lawn at inception, they will make their presence known in due course. High rank is not specifically or irrevocably fixed to a species since different species will behave differently in different conditions; it is a matter of homegrown judgement but would include creeping buttercups and white clover in lawns that suit them well.

SUMMARY

- Use many less vigorous species and in greater quantities and only few locally vigorous ones.

PLANT SIZE

Many of the more commonly available ornamental plants that are suitable for use in tapestry lawns such as *Ajuga reptans* (bugle) and its cultivars, can be sourced as individuals in a variety of different sizes, ranging from extra value, standard and jumbo plug plants to 9 cm and 1.0 L pots. It is yet another of the many peculiarities of current commercial horticultural practice that plug sizes are not generally standardised and will vary between producers, and that most pot sizes are referred to by their volume capacity in litres with the general exception of pots that are smaller than 1 L, which tend to be referred to by the diameter of the top of the pot.

Plug plants of any size are juvenile plants that may have been grown vegetatively from cuttings (which is most likely if they are cultivars) or from seed. The type of roots that the plug plant will have developed can be different depending on its propagation material. The roots of plants grown from seed will have originally developed from a single initial root known as a radicle. These initial roots tend to be relatively strong, and in many dicots can develop into relatively deep-running and carbohydrate-storing tap roots (A) (Figure 2.38).

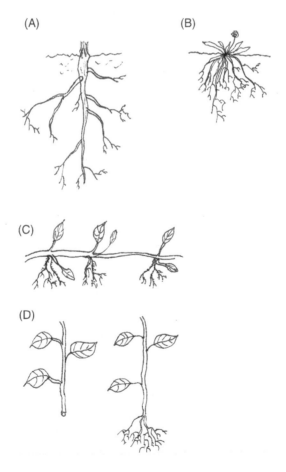

FIGURE 2.38 Plant roots. (A) Taproot, (B) fibrous root, (C) adventitious root and (D) adventitious roots developing on a cutting.

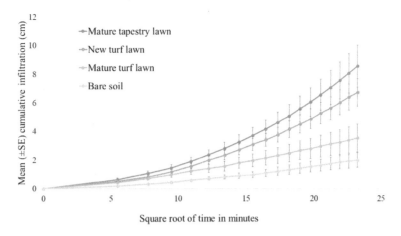

FIGURE 2.39   Water infiltration in tapestry and turf lawns on a silty loam soil suggested that in lawns over three years old that greater water infiltration rates could be found in T-lawns. Grass lawns generally develop thatch and have a shallower rooting zone densely populated by fibrous roots from only a few grass species, while T-lawns do not develop thatch and have a diversely populated root zone with many deeper rooting species.

Plants grown from cuttings develop adventitious roots, which are not as initially robust or as deep rooting (C+D). This becomes relevant if you are interested in the subsurface characteristics of your lawn, since continuous stretches of shallow fibrous roots (B), such as found in traditional grass lawns, appear to delay water infiltration when compared to T-lawns (Figure 2.39). There are a variety of factors that influence drainage. However, since T-lawns are only lightly managed and therefore less walked upon and made with plants with a variety of rooting depths and forms, it seems plausible that these features may be part of the explanation. The size and number of roots can also influence plant survival and regeneration post-mowing due to their carbohydrate storage capacity.

   A. Taproot.
   B. Fibrous root.
   C. Adventitious root.
   D. Adventitious roots developing on a cutting.

PLANT TILES

Unless the lawn happens to be small, the most aesthetically pleasing results have so far not been produced by using individual plants as plugs or in pots, but rather by using single plant species grown in 3 L British standard seed trays (344 × 214 × 52 mm), sometimes referred to as 'flats' in other parts of the world. The unusual metric measures derived from the old imperial 13.5″ × 8.5″ × 2″ seed trays. Half-sized seed trays can also be used, although the visual impact is reduced and successful establishment may be less certain.

By using seed trays rather than pots, something akin to a size-standardised patch of plants is produced, something that for all intents and purposes can be regarded as a 'plant tile' (Figure 2.40). The tiles can then be laid in a checkerboard-type pattern, creating instant patches of vegetation of a size that is immediately distinguishable to the human eye, whereas when using smaller plug plants and pots the individual plants can be overwhelmed by their neighbours both visually and via competition between species. Over time, clonal plants used in this way will spread and intermingle with their neighbours (Figure 2.41), while non-clonal plants will largely remain as patches where they were planted.

FIGURE 2.40   A well-rooted single species plant tile of *Chamaemelum nobile* 'Flore Pleno' ready for laying.

FIGURE 2.41   Although originally planted in tray-sized patches, most T-lawn plants will spread, creep and mingle. Shown here 3 years after planting, a 10×10 cm^2 of T-lawn contains 12 visible species with *Achillea millefolium*, *Ajuga reptans* 'Atropurpurea', *Bellis perennis*, *Chamaemelum nobile*, *Glechoma hederacea* 'Variegata', *Leptinella squalida* 'Platt's Black', *Mazus reptans*, *Parochetus communis*, *Prunella vulgaris*, *Ranunculus repens*, *Veronica chamaedrys* and *Veronica officinalis*. This mixing will affect the overall patch-type aesthetics; however, unlike traditional well-spaced garden planting, it compares not unfavourably with the up to 25 wild unselected species per 10×10 cm^2 found in the much taller-growing vegetation of Europe's richest meadows. (From Kull, K. and M. Zobel, *J. Veg. Sci.*, 2, 715–718, 1991.)

COMPETITION BETWEEN PLANT TILES

Plants compete in many ways both above and below ground, and one of the areas of strong competition is at the juncture where individual plants meet each other. They jostle and compete for space, light, water, nutrients and the like. This can be between related plants such as clones and siblings, and other plants of the same species, such as those that might be grown together as a single species group in trays to make a plant tile and is known as intraspecific competition (Figure 2.42). When the competition is between plants of different species, such as that between different species plant tiles, it is known as interspecific competition. Of the two types, intraspecific competition in plants is usually thought to be more intense since plants of the same species compete for the same resources; however, there is also evidence that directly related individuals of the same species may not compete as intensely with each other as with unrelated individuals [41]. Also, of relevance here is that in the propagating environment of a single species seed tray all things are not competitively equal, and those plants at the edge of a tray may be better developed as they face less direct competition and have greater access to space and light, something often referred to as 'the edge effect'.

Observation of newly laid T-lawns suggests that for many, but not all, of the species used in T-lawns, that plants at the centre of the tile appear to establish themselves somewhat faster after laying than those plants at the edges. They appear to catch up in their development. Although there are too many variables to offer an explicit explanation of this behaviour, the type and degree of competition seems likely to be at least partly influential. This behaviour tends to be most apparent in relatively larger species, less so with small-leaved ground-hugging ones.

Bearing this in mind, laying of lawns has been adapted from an entirely random planting methodology to what might be thought of as graded planting based loosely on plant morphology and growth habit, with trays of small-leaved ground-hugging plant types being planted together in 'similarity clusters' and the relatively larger species filling the gaps between the clusters (Figure 2.43). This does not guarantee outcomes but appears to facilitate good establishment across the lawn.

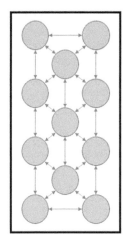

Intraspecific competition
within a plant tile.

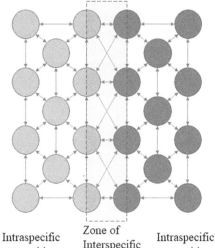

Intraspecific
competition

Zone of
Interspecific
competition

Intraspecific
competition

FIGURE 2.42   Initially the central area of the plant tile on the left is under the most competitive stress from members of its own species (shown in dark green). Plants at the edge of the tile may be better developed as they face less direct competition and have greater access to light, something often referred to as 'the edge effect'. Once planted the usually better developed edge plants are then subject to interspecific competition with neighbouring species (shown in blue). While the edge plants of both species engage in both intraspecific and interspecific competition, the plants in the centre of the tray remain intraspecifically competitive only.

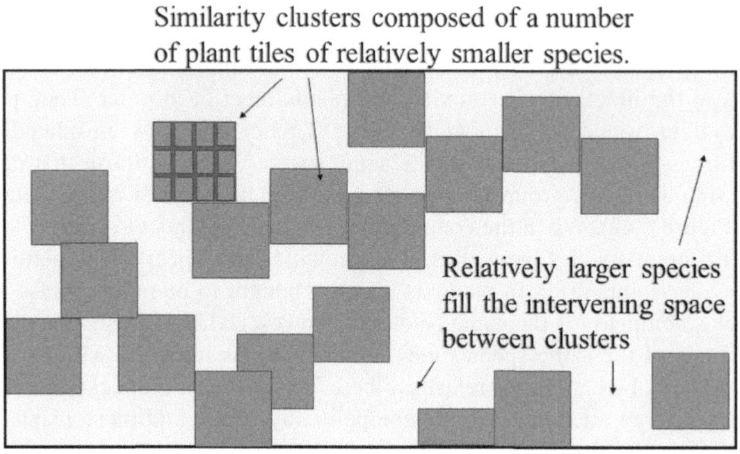

Similarity clusters composed of a number
of plant tiles of relatively smaller species.

Relatively larger species
fill the intervening space
between clusters

FIGURE 2.43 Planting a T-lawn with clusters of low-growing and ground-hugging species embedded amid the relatively larger species has been effective in facilitating establishment. However, it can lead to what appears to be low path through the lawn and induced footfall may follow.

If the plants were not in tray-sized tiles and were instead individuals, the individual would be both trying to establish and compete with the unrelated species surrounding them at the same time. Amongst other factors, the size and relative vigour of the individual plant and its rate of spread would then be particularly relevant; is it a daisy or a creeping buttercup for example. A larger or more vigorous plant is likely to have the advantage and potentially overwhelm its neighbours. With the provision of preformed patch-sized plant tiles, it seems less likely that the new neighbours will be able to overwhelm the patch before its centre has time to establish.

Although this is at best a generalisation, single species plant tiles have proven to be almost always successful at establishing and subsequently persisting for more than at least one growing season when used in this way, while plants planted using an alternative method such as single plug plants or as single pots (unless they are very vigorous species such as creeping buttercups) have lower successful establishment rates over the same period.

Using plant tile patches for most species is usually appropriate and effective. Single pots of the locally most vigorous species (rank 1–4) can then be planted within the tile patchwork; hopefully this sparse planting will then act to ameliorate any competitive advantage that they may ordinarily exhibit.

LAWNS FROM A SEED MIX

There has been much interest in producing T-lawns from seed, ideally a simple scattered seed mix that will produce a T-lawn without the requirement for plants in trays or pots. To date, this has proven to be a less than ideal method and does not lead to the most colourful and refined style of tapestry lawn as outlined.

WHY NOT A T-LAWN FROM A SEED MIX?

It is indeed possible to put seed from most of the useful species into a single packet (although a few are very difficult to obtain commercially or in useful quantities), but the outcome is very different from that obtained via plant tiles and the use of coloured foliage cultivars.

In a seed-mix, there will be very fine and difficult-to-harvest seed, for example, *Lobelia angulata*, which tends to be expensive and takes up little space within the mix, and relatively large and easy-to-harvest seed, for example, *Trifolium repens*, which tends to be cheaper and takes up more space within the mix. Larger seeds produce larger seedlings, and these have the capacity to outcompete any nearby small seedlings. The larger seeded and generally cheaper species can quite simply

FIGURE 2.44   1.5 m² trial lawns using seed mixes. The four rear lawns use mixed seed from 24 species in appropriate proportions. The front two lawns shown were sown at the same time using the same species but applied in discrete patches. The patches of visible substrate indicate where species with small seedlings, or that are slower to germinate, have been sown.

overwhelm the smaller more expensive ones. To ensure that at least some of the species with expensive smaller seed manage to survive, they must be included in the seed mix in very large quantities, substantially increasing the cost.

Experiments with seed mixes have all produced less-than-ideal results (Figures 2.44 through 2.46). In what might be considered as the best outcome for a seed mix so far, a mix of over 40 different species resulted in a lawn with only 12 constituent species after 2 years of growth, and two of these were errant species not originally included in the seed mix.

There is also a concern regarding obtaining seed of the correct phenotype. Seed companies and their suppliers are in the business of selling seed; in general, for wildflower-type mixes, they tend to

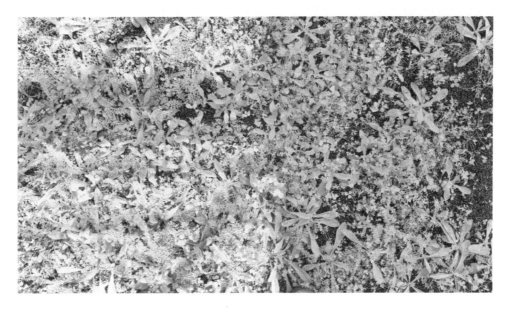

FIGURE 2.45   A closer view of a developing mixed seed lawn. Such a lawn excludes vegetatively propagated cultivars. It is dominated by quickly germinating and swiftly growing species.

FIGURE 2.46   A closer view of a developing patch-sown lawn. The variation in young plant size, variability in germination and subsequent plant development can be clearly seen. Bare patches contain seed yet to germinate.

grow wild species phenotypes that produce commercially viable quantities of seed and are relatively easy to harvest. There is nothing wrong or underhanded in this, but a tall-growing wild phenotype can be easier to harvest from than a prostrate phenotype and may produce a greater quantity of seed. It is possible to order seeds of plant species X and receive seeds of plant species X, but when grown, species X may not be the prostrate or decumbent phenotype hoped for. Imagine ordering *Trifolium repens* seed and receiving the ladino cultivar. The same can sometimes occur with *Lotus corniculatus, Mentha pulegium, Prunella vulgaris* and *Trifolium pratense* to name but a few. It is very wise indeed to check before ordering wildflower seeds for a T-lawn that the phenotype you will receive is a low-growing rather than erect form and is therefore more suitable for use in a T-lawn.

Furthermore, some species' seed require a period of cold before they will germinate, for example, *Viola* sp. Unless they have been pre-treated (which will increase the price), this type of seed is unlikely to germinate without sufficient chilling; they may not show for a year if they are first sown in spring. With sufficient chilling, seeds that survive the year unmolested are likely to germinate, but they will be starting life within an established lawn and will therefore be at a competitive disadvantage. Also, already established lawns will most likely be home to and receive visits from a variety of both seed and seedling munchers.

Additionally, some seed is also notorious for erratic germination, for example, *Acaena* sp., or it may require specific environmental stimuli, with perhaps some of the seed easily germinating on sowing, while some may germinate a few months later and some may lie dormant for even longer. This is a survival strategy used by some plants, not necessarily a problem with the seed, but it has so far proved too challenging to contain all these types in a 'simple' seed mix. A dicot-rich lawn of swiftly germinating seed types is certainly possible by this method, but it does not produce a recognisable patchwork-type tapestry lawn.

OVERSOWING WITH SEED

Although not a species usually requiring chilling, and with relatively large seed that can germinate quite quickly, experimental oversowing of *Viola tricolor* (heartsease) seeds into established tapestry lawns suggest that seedling success rate can be around one adult plant for between every 50 to 70 seeds sown. A similar outcome has been achieved with ornamental *Bellis perennis* (daisy). The outcome of oversowing will inevitably be dependent on the species used and the size and robustness of the resulting seedling [42], but if this type of result is at all indicative of seedling success in an established T-lawn, then it seems likely that a large quantity of seed is needed for a good chance of success.

The exclusion of vegetatively propagated cultivars with colourful leaves, for example, *Lysimachia nummularia* 'Aurea' (golden creeping jenny) and *Ajuga reptans* 'Burgundy Glow' (variegated bugle), results in seed-sown T-lawns that are entirely green in colour. They also do not exhibit the visually interesting patchwork or drift-type effect produced by using plant tiles but instead are more homogenous looking, especially from a distance. Although longer-term experiments have yet to be undertaken and a simple mixed green foliage T-lawn may be possible, a T-lawn seed-only mix that produces the same colourful foliage and flower tapestry of a plant-tile-created lawn seems unlikely any time soon.

3 Species Roles in Tapestry Lawns

A look at the behaviour of the types of plants that can adapt successfully to the lawn environment shows that not all are superbly floriferous. Many have a seasonally limited floral period producing flowers mainly in spring and early summer, and some have rather small (for example, *Lobelia pedunculata*), inconspicuous (for example, *Clinopodium douglasii*) or surprisingly few flowers (for example, *Potentilla reptans*).

From a human viewpoint, many of these species do not substantially add to the overall floral performance unless they are clearly seen in very large numbers (which is possible with the small-flowered *Lobelia* sp.), and the most impressive floral show will tend to come from those plant species with the largest and therefore most visible of flowers [43]. There is a temptation here to select plants that are the most florally impressive; however, this is a managed groundcover mixed-plant community and despite its obvious attraction, floral performance is really the icing on the cake.

Plants with little floral value but that give good ground cover are essential, as is their contribution to the overall number of species. It is also worth considering the value of scented or coloured foliage, and perhaps if ecologically minded, which species are useful food or nectar plants for pollinators, or how lawn plants such as *Trifolium repens* might contribute to the robustness of the community, with its mowing related and therefore potentially cyclical addition of small amounts of organic nitrogen from the nitrogen-fixing bacteria associated with its roots. T-lawns may function best without supplementary fertiliser, but they do require nutrition.

All T-lawn plants are part of the overall ground cover effect at some time in the year; however, they can be broadly classified into three categories:

- Providing a conspicuously floral contribution. These plants tend to have relatively large flowers or clustered flowerheads, for example, *Trifolium repens*, or produce large numbers of small flowers, for example, *Lobelia pedunculata* (pratia).
- Providing a primarily vegetative ground-covering role. These species tend to have inconspicuous or few flowers, for example, *Leptinella dioica*.
- Providing a combination in part of the two but not being especially floriferous or especially ground covering, for example, *Potentilla reptans*.

A lawn with poor ground coverage is not a lawn at all, and a tapestry lawn is most effective in its floriferous, species-rich and ground-covering role when it contains a mix of all three types in broadly similar proportions. Around one-third of notably floral plants, one-third providing mostly vegetative ground cover and one-third that can contribute both to some degree. Absolute thirds are not a strict requirement here but rather a starting guideline for an initially balanced type lawn with good ground cover and the potential for an eye-catching floral show.

LEAF COLOUR

Wild-type plants rarely have mature leaf colour beyond various shades of green; it is the most photosynthetically relevant colour since it generally means that the leaf is being efficient in its use of light. Healthy deep-green leaves tend to contain greater amounts of chlorophyll than paler leaves

FIGURE 3.1   With only a few disappearing daisies offering flowers, it is the mix of leaf colours and variegation in a mostly shady part of a lawn that provides the colourful visual interest not found in traditional lawns.

and are usually the most productive. Wild-form plants such as the creeping buttercup typically have such chlorophyll-rich leaves; it helps give them the edge over competitors. However, visually interesting leaf colour variations are available via the use of selected cultivars. These have other prevalent pigments and can add another level of visual interest to an otherwise green lawn. The use of coloured-leaved cultivars is usually a particularly admired feature of T-lawns since it can provide colour in the lawn throughout the year and is invaluable for the periods when few plants are in flower (Figure 3.1).

The most frequently encountered cultivar leaf pigmentations tend to be shades of red and yellow. Red leaves tend to have greater amounts of red pigments such as anthocyanin than plain green leaves but generally function in much the same way; the green associated with chlorophyll is frequently masked by the greater amounts of red pigment. Red and purple pigments are also thought to have protective properties and can act a bit like sunscreen and antifreeze, which is why some young leaves in spring and frost-touched leaves in winter can have reddish or purplish pigmentation. Some red pigments don't taste nice either and can deter munching insects, but not necessarily larger herbivores. Wood pigeons for example (with substantially fewer taste buds than us humans) can see the world in a colour spectrum similar to our own, and red-leaved clovers stand out against a generally green lawn background like cherries on a cake. The little blighters have been observed selectively plucking and consuming every red clover leaf in a lawn while leaving the green leaves virtually untouched until their next visit (Figure 3.2). Biodiversity can make you wince just a little sometimes.

Yellowish-leaved cultivars tend to show a different set of leaf pigment such as those produced by a pigment group known as the xanthophylls, but rather than masking chlorophyll in the manner of red leaves they usually indicate that chlorophyll and other pigment levels are lower than in normal

FIGURE 3.2   After a visit from wood pigeons, just the petioles of this red-leaved white clover cultivar remain.

green leaves; the leaf is likely to be less efficient than a fully green counterpart. The same applies to most yellow or white-variegated leaves. Yellow-leaved cultivars are therefore usually less vigorous plants. In a natural environment they would be out-competed quite quickly. 'Usually' since sometimes the lesser vigour is not always readily apparent. The yellow-leaved creeping buttercup cultivar 'Buttered Popcorn' is generally not as vigorous as its green-leaved counterpart, but given ideal conditions it is sometimes hard to tell the difference. In T-lawns where nutrient levels are hopefully a bit lower than in cultivated soils, the difference in vigour is a bit more noticeable, and *R. repens* 'Buttered Popcorn' is usually both better behaved and a worthwhile colourful addition.

Sometimes the variegated form of a plant can be substantially weaker than the unvariegated form and may be outcompeted if both are included in the lawn. *Glechoma hederacea* 'Variegata' (variegated ground ivy) can slowly dwindle in coverage if the unvariegated form is also in the lawn. It may be best to choose one form or cultivar and stick with it. It is also possible for variegated plants to revert to their non-variegated form.

Although not due to pigmentation but rather a covering of fine hairs, silver/grey can be encountered in T-lawns. It is a feature that tends to be associated with dryish, warm and sunny conditions, as the fine hairs can act to protect leaves from intense sunlight and water loss. If you have these conditions, you might like to include *Argentina anserina* (silverweed) or *Pilosella tardans* (Syn. *Hieracium pilosella* var. *niveum*—silver-leaved mouse-ear hawkweed) in your plant mix. A bit of silver and gold in a lawn can be quite effective (Figure 3.3).

A bit of horticultural foresight when planting can produce some delightful combinations of leaf and flower, with leaves acting to provide a contrasting or complimentary background in the manner of traditional garden horticulture (Figure 3.4), although this careful type of planting tends to be short-lived once the plants begin their self-organisation in response to the conditions. It is important to bear in mind that this is a mobile plant community and not a fixed planted border; nothing is going to stay the same for long and there are no guaranteed outcomes.

FIGURE 3.3 *Argentina anserina* and *R. repens* 'Buttered Popcorn' adding a touch of gold and silver to a lawn. Alternatively, it is possible to use simply *Argentina anserina* 'Golden Treasure', a cultivar with variable gold variegation through many of its leaves.

FIGURE 3.4 Coloured leaves can provide contrast and act as a foil to accent the short-lived flowers. Here a simple late autumnal mix of *Acaena inermis* 'purpurea', *Fragaria vesca* 'Golden Alexandria' and *Parochetus communis* is quite effective. *F. vesca* 'Golden Alexandria' is not a clonal strawberry that spreads by runners and can be a bit pernickety in T-lawns and is therefore not on the recommended list, but here it demonstrates how a simple mix of leaf colours can be very attractive.

FIGURE 3.5   Come the warmer temperatures at the start of spring the winter-dark leaves of *Lysimachia nummularia* (right) are pushed aside and discarded by the plant for the more familiar fresh lime-gold leaves of this common hanging basket cultivar. *R. repens* 'Buttered Popcorn' (left) also shows the first hints of the variable yellow leaves that are to come. *Ajuga* 'Chocolate Chip' is just getting going between them.

Seasonality can also have an influence on lawn leaf colour. The most vibrant colours tend to occur in spring on newly produced leaves (Figure 3.5), while later in the year autumn provides shades of golds and browns as the deciduous plants such as *Argentina anserina* die back for the winter (Figure 3.6). Some T-lawn plants will respond to the falling temperatures by taking on a darker colouration as protective pigments come into play (Figures 3.7 and 3.8).

FIGURE 3.6   No flowers, only foliage. As autumn arrives, hints of brown and gold add to the multitude of foliage colours.

FIGURE 3.7 Golden-leaved *Lysimachia nummularia* (creeping Jenny) showing a complete change from theusual bright-yellow leaves as it responds to lower temperatures.

FIGURE 3.8 Touched by dropping temperatures, the leaves of this *Veronica officinalis* (heath or commonspeedwell) have darkened as protective pigments come into play.

Compared to the busy growing season's ever-changing mix of bright leaf and floral colours, a winter T-lawn can have a still and darker, sometimes almost brooding, character of its own.

The overall effect of the rich variations in foliage size, shape and colour is to draw the eye in to the construct itself and seek out the finer details rather than just focus on the flowers that it may contain. Not many horticultural constructs draw you to observe the small details; T-lawns most certainly can. Once drawn in the eye tends to wander, often observing without clear purpose, just looking. A surprising number of people have been seen standing still and staring into tapestry lawns in a contemplative and almost meditative manner (your author included). On questioning, the majority of T-lawn visitors report they are enchanted by what they see; although some are wondering what they are looking at and a few seem unable to put their finger on what is different about the lawn until it's pointed out to them. The latter group may offer a type of positive affirmation of the T-lawn being a 'real' lawn since they notice no real difference, or perhaps it indicates that for some people details like the composition of a lawn are just not the sort of thing that they would usually notice.

SCENT

There is yet another feature of T-lawns to consider, a sensory one that harks back to the earliest medieval lawns, that of scent. Although it is possible to find scent in the flowers that can be incorporated in T-lawns, it is the scent of bruised foliage that is by far the most noticeable; a scent that develops when the plants are walked upon and when they are mown.

Most people will be familiar with the scent of mown grass, the scent of a mown T-lawn is most definitely different. There is inevitably the scent of cut herbage, but captured within it are the mixed fragrances of all the leaf-scented constituent plants and depending on those constituents the bouquet can be strongly aromatic. A gathering of noses down wind of the lawn as it is mown will almost surely be sensitive to it. Young children in particular seem to be drawn to the scents in T-lawns, as they almost unhesitatingly, and often with great enthusiasm, trample and crawl over them eager to find the best smells, and any four-leaved clovers that might be found; the canny lawn owner having included *Trifolium repens* 'Pentaphyllum' (Syns. *T. repens atropurpureum* and *T. repens purpurescens quadrifolium*) for just such a purpose. Children seem to find sitting on flower-filled T-lawns and poking and prodding and picking the daisies to be a most natural and enjoyable thing.

The bubble-gum spearmint scent (and taste) of *Micromeria douglasii* (Indian mint) is always a favourite, and the coarse peppermint of *Mentha pulegium* (creeping pennyroyal) can make a nose curl. *Chamaemelum nobile* (Roman chamomile) is another favourite that is difficult to have too much of in a T-lawn, so long as it is the creeping variety, and *Achillea millefolium* can add a similar-type scent, although one of its rarer bygone names 'Sneezewort' can have a new relevance for some people. Some have said the same about *Glechoma hederacea*, which also has a distinctive scent. If your lawn is in full sun and on the dry side, then *Thymus praecox* is almost a must have, although it must be a creeping species, as so far, the other trialled thymes tend to be slow to regenerate, become woody and generally don't take well to being mown.

FOOT TRAFFIC

T-lawns can easily withstand young knees and a surprising amount of foot traffic. They may even benefit from being walked upon as the plants are pressed down, shaped and stolons are kept in contact with the soil. So long as footfall isn't repeated in the same places too frequently, T-lawns should show little long-term damage, but it is worth remembering that forb leaves are not the narrow silica-reinforced leaf blades that distinguish grasses. T-lawns are not designed with regular footfall in mind; you might consider stepping-stones or pathways if you need repeated access via your T-lawn, and if ball games are your thing, then it is probably best to stick with a traditional grass lawn.

Choosing plants for footfall is challenging, since people will walk where they may and will often follow desire lines (the shortest route between A and B) rather than any direction you might prefer them to take. People also tend to follow anything that resembles a path, and in T-lawns paths appear to exist in areas with mostly prostrate plants. You may find visitors repeatedly walking on the prostrate plants without specifically meaning to, simply because walking on taller plants isn't something generally done. If you are determined to have a living plant path rather than using stepping-stones, gravel or mulch-type paths, it is useful to include daisies and yarrow, as they can stand up to a bit of wear and tear, but it is also helpful to be the one who pre-treads the path to determine its route. If you walk your path from time to time and keep it defined, others who walk your lawn will tend to follow it.

Unlike young children, adults can be initially quite reticent to step onto a tapestry lawn. Perhaps it is an assumed or conditioned behaviour; after all, who walks on flowers and isn't a vandal? However, you can remind them that daisies, buttercups and clovers all inhabit traditional grass lawns, and we don't have a problem walking on them there. It just takes a few bold steps.

SUMMARY

- Useful T-lawn species fit three general categories: primarily floral, primarily vegetative ground cover and intermediate combinations.
- Broadly equal proportions of each category are a good start for new T-lawns.
- The use of cultivars with coloured leaves is a particular feature of T-lawns.
- Red-leaved cultivars behave much the same as their green-leaved forms, while yellow-leaved cultivars tend to be less vigorous.
- Variety in leaf shape, size and colouration is as important as floral performance.
- Leaf scent is well-worth considering when selecting plants.
- T-lawns can withstand light and non-regular foot traffic.

4 Wildlife

Monoculture-type grass lawns are not renowned for the rich variety of wildlife that they can support. Above ground, a few grass species mown on a weekly basis do not provide an ideal habitat. Such a low-cut grass lawn devoid of forbs has essentially no value to nectar-feeding pollinators, since if in flower, grasses are wind pollinated and of limited value to grass-munching invertebrates, and in turn those that feed on them; it is perhaps no surprise to learn they have been called 'green deserts' [44].

Tapestry lawns are not monocultures, neither are they devoid of forbs, and with mowing at between three to five times a year they are disturbed far less frequently than traditional lawns. A consequence of this is that T-lawns can contain and support a surprising variety of insects and other invertebrates. These are important in themselves but also because they are low down on the food chain and an integral part of the complex food webs that gardens can support.

INVERTEBRATES IN THE LAWN

An early study into the small invertebrates that might inhabit T-lawns composed of only 10 different forb species had some interesting results. The kind of invertebrates to be found varied widely from woodlice, centipedes and molluscs to moths, sap suckers, spiders and all manner of little creepy crawlies, with over 80 different invertebrate families being recorded (Figure 4.1).

Before revealing the outcome of the study, it is worth mentioning that to allow for a comparison between T-lawns and grass lawns, that the comparative grass lawns grown for the investigation were cut only when the T-lawns were cut, that is, not weekly. This is not the usual frequency for grass lawns at all. Grass lawns would normally be cut much more frequently, about three time more frequently than they actually were, so the level of disturbance they received is substantially lower than would occur normally, and the grass had the time to grow much taller than would be expected of a typical garden lawn. A consequence of this is that we would expect to find more invertebrates in the experimental grass lawns than would normally be there under a more usual lawn-mowing regime.

Firstly, and unsurprisingly, the taller the lawns grew before mowing, the more invertebrate life they contained, and the shorter they were kept by the mowing regime the fewer the number of invertebrates. This seems to be a regular feature of ornamental plantings; it's that bungalow and multi-storey building scenario again [45]. With more vertical lawn space to inhabit, more invertebrates could be found within it, and this applied to both tapestry lawns and grass lawns. However, lawns are necessarily height restricted, and as previously outlined, tall growing T-lawns can pose problems for plant species survival and diversity, so the results mentioned here are taken from the applied mowing height of 4 cm only; the T-lawns being cut back to 4 cm once they had reached 6 cm.

With this mowing regime, the second notable outcome was that the number of invertebrates and the overall diversity within the invertebrate families studied was broadly equivalent in both grass and T-lawns. A bit of a surprise perhaps but remember that the grass lawns were being cut with the same frequency as the T-lawns and consequently had been allowed to get substantially taller than would be usual (in fact they grew much taller than the T-lawns), and were much less disturbed than would be usual.

Unless the beasties collected in the study had a fear of heights, that the lower-growing 10 species T-lawns contained both similar numbers and diversity of insects prior to mowing would suggest that the concentration of invertebrate life was greater in the substantially lower growing T-lawns. This was particularly of note since surrounding the experimental grounds were several hectares of

FIGURE 4.1 Spot the spider. Look amid the *Thymus praecox, Pilosella officinarum, Argentina anserina, Erodium castellanum* and *Viola riviniana* for a splash of gold foliage.

managed grass lawns and grassland-type meadows, and these could be a feeder source for the grass lawns, while there were only a few small experimental T-lawns nearby.

Both the T-lawns and grass lawns shared a surprising amount of invertebrate species but also had distinct population differences, meaning that both lawn types support different types of invertebrate community. This is relevant if you consider the vast swathes of grass lawn that exist across the country and the inevitable similarity of the invertebrate community type that can be found there; the T-lawns fostered a significantly different type of lawn community and therefore added to overall landscape biodiversity. The study also showed no negative effects on this diversity if non-native plant species were included at 50% in the T-lawns, although non-native-only T-lawns did not fare as well and invertebrate diversity was found to be significantly lower [46].

What this early study did not manage to provide was a direct comparison between plant species-rich T-lawns cut between three and five times a year and grass lawn cut with a more typical frequency of between 20 to 30 times. However, that a T-lawn of just 10 plant species can hold the same amount of invertebrate life as a traditional grass-only lawn and add to overall landscape biodiversity is strongly suggestive that species-rich T-lawns with many more plant species are likely to be richer in invertebrate numbers and variety than their grass counterparts if grown and mown according to their usual T-lawn format requirements.

BEES, HOVERFLIES AND BUTTERFLIES

The urban and suburban environment, where these days most people tend to live, may not seem the ideal place for winged pollinators, but green-fingered inhabitants of towns and cities have the capacity to play an important role in ensuring insect pollinators survive and thrive around them [47]. Tapestry lawns may be one of the methods to assist in this.

A closer look at how three groups of pollinators interacted with plant species-rich T-lawns revealed that tapestry lawns are significantly more attractive to bees, hoverflies and butterflies, and by a substantial margin, when compared with a pure grass lawn or a commercially available 'flower lawn' that contained grasses and nine different forbs (Figure 4.5).

'Flower lawns', 'Wildflower turf' and 'Meadow turf' tend to be mixtures of grasses and forbs in varying proportions, usually allowed to grow substantially taller than traditional height-constrained lawns, and they may have a similar or less frequent mowing requirement to T-lawns (Figures 4.2 through 4.4). Dependent on type, they are mown usually twice a year and are closer in form and management to meadow-style planting. 'Wildflower-enriched lawns' retain the traditional grasses

FIGURE 4.2 A strip of wildflower turf grown as a border to a wall. Wildflower turf is usually well-endowed with a variety of forb species and is not meant to receive the usual frequency of mowing expected for a traditional lawn, particularly since many of the flowering species require height to perform at their floral best.

FIGURE 4.3 An example of meadow turf in early spring. For the most part, a different set of species to that used in T-lawns is used in meadow formats and the most floral period tends to be later in the year.

FIGURE 4.4   An example of meadow turf in high summer. It is seasonally flower-rich; the height is unrestricted unlike that of a tapestry lawn and can attain over 1 m.

and regular mowing but purposefully include mixed forb species in greater proportion than is usual, and mowing frequency may vary over time in line with the season.

It might seem obvious that flowering, species-rich T-lawns would be more attractive to pollinators than grass-only lawns, but until someone did the legwork and looked, it remained supposition. To get a better understanding of any differences, an experiment monitored pollinator activity over a period of three summer months on grass-only lawns, the commercially available 'flower lawn' format and an early forb-only T-lawn format. The results showed that all lawn types received visits from pollinators, although it was noted that on grass-only lawns that the pollinators were resting rather than doing any foraging. By the end of the study it had been observed that for every 1 visit by a pollinator to grass-only lawns there were over 8 visits to the flower lawns and over 80 to the tapestry lawns. Not only were tapestry lawns found to be visited significantly more frequently than the other lawn types, but the variety of visiting pollinator species was significantly greater too. Nine pollinator species were recorded as resting on grass-only lawns, 25 species observed to visit the flower lawns while 45 species were recorded visiting tapestry lawns. Overall, that's lots more bees, hoverflies and butterflies visiting T-lawns a lot more frequently than the other two formats (Figure 4.5).

Inevitably the location, timing, size and plant components of T-lawns will influence which pollinators visit them, and how frequently, but it is now clear that species-rich T-lawns can be very attractive to a wide range of pollinating bees, hoverflies and butterflies (Figures 4.6 and 4.7).

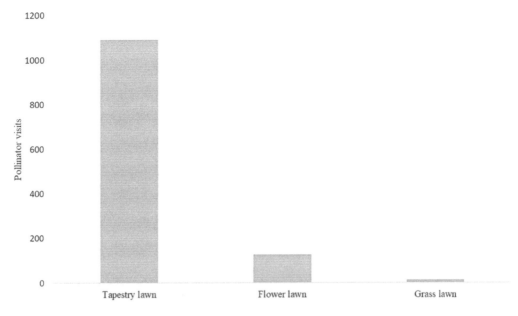

FIGURE 4.5   Pollinator visits. The attractiveness of each of the lawn formats can be demonstrated by the number of pollinator visits. In trials over three summer months, for every 1 visit by a pollinator to a grass lawn over 8 visits were recorded on flower lawns and over 80 on tapestry lawns.

FIGURE 4.6   A painted lady butterfly (*Vanessa cardui*) sipping nectar from *Lobelia pedunculata* in a tapestry lawn.

FIGURE 4.7 *Trifolium pratense* was the most frequently visited flowering species in T-lawns during the three-month study period.

Other pollinators such as day-flying moths (Figure 4.8), flies and a multitude of flying insects can also be seen on T-lawns. What pollinators may visit after dark has yet to be investigated.

During the study it became apparent that certain plant species received more visits from pollinators than others. There will be many factors to influence this such as (a) seasonality, since some plant species were past their floral peak or not in flower due to the season; (b) the pollinator species

FIGURE 4.8 An adult, day-flying, six-spot burnet moth (*Zygaena filipendulae*) feeding on lawn *Achillea millefolium* flowers. As a caterpillar, it would have fed on another T-lawn plant species *Lotus corniculatus*.

themselves, since some pollinator species have floral preferences, for example, for open flat-headed flowers, and some exhibit floral constancy sticking to visiting one plant species at a time; (c) flower number, since some plants produce few or irregularly timed flowers; (d) flower size, since larger floral heads tend to accommodate more pollinator activity and (e) mowing, since post-mowing flower numbers are significantly reduced. However, for those interested in selecting plant species with pollinators in mind, the 14 most visited lawn plants during the three summer months of study are as follows:

1. *Trifolium pratense* (red clover)
2. *Pilosella aurantiaca* (fox & cubs)
3. *Prunella vulgaris* (selfheal)
4. *Thymus praecox* (mother of thyme)
5. *Chamaemelum nobile* 'Flore Pleno' (double chamomile)
6. *Potentilla reptans* (cinquefoil)
7. *Lobelia pedunculata* (pratia)
8. *Ranunculus repens* (creeping buttercup)
9. *Trifolium repens* (white clover)
10. *Argentina anserina* (silverweed)
11. *Achillea millefolium* (yarrow)
12. *Bellis perennis* (daisy)
13. *Lotus corniculatus* (bird's-foot trefoil)
14. *Campanula rotundifolia* (harebell)

This list should not be regarded as a definitive list of the 'best' plants to use, rather as a simple indicator that during summer these particular species can be notably attractive to pollinators when included in T-lawns.

Intriguingly, those plants that occurred in both tapestry lawns and flower lawns saw significantly different numbers of visits. For example, *T. pratense* in T-lawns was visited over 30 times more than *T. pratense* in neighbouring flower lawns that were 1.5 m away. Without further investigation it is difficult to be sure why this might be since flower numbers in each format over the three-month period were variable. However, it suggests that the T-lawn format of low-cut, easily accessible and concentrated floral resources may well be influencing pollinator choices and activity.

Not only invertebrates can make use of tapestry lawns. Young amphibians have repeatedly been observed using the structural complexity of T-lawns for shelter, taking refuge in the amenable conditions that can be found there, and no doubt snacking on the lawn invertebrates (Figure 4.9). Unfortunately, they tend to be driven from the lawn by mowing when it occurs, but if they should linger, then so long as they stay below the 4 cm cut height when the mower is passing over and can avoid being squished by a misplaced boot they seem none the worse for the brief experience.

Hedgehogs (*Erinaceus europaeus*), which tend to show a preference for less dense and generally shorter foliage than is found in tall meadows or grasslands, have been observed nosing through T-lawns, and blackbirds (*Turdus merula*) and other ground-foraging birds in search of seeds and invertebrates can be frequently seen attending to T-lawns. Wood pigeons (*Columba palumbus*) seem particularly fond of them. This is probably due to the inclusion of clover species in the plant mix, since during winter in particular, clovers can make up over 50% of a wood pigeon's diet. They have an annoying habit of picking at flower buds too, but they are not alone in that. Rabbits can be drawn to T-lawns and can indulge in some perplexing behaviour, munching on plants that they seem largely to ignore in other neighbouring locations (spring bulb flower buds in particular), and digging little latrines in new and relatively new T-lawns. Fortunately, this is a habit they seem to indulge in less as the lawn matures.

The inclusion of early-flowering geophytes such as snowdrops, *Scilla sibirica* and *Crocus* sp. can also be attractive to pollinators that become active on warm days early in the year (Figure 4.10).

FIGURE 4.9   Disturbed by mowing, a young common toad (*Bufo bufo*) emerges from what was a cool, moist and leaf-shaded part of the lawn.

FIGURE 4.10   Brought out of hibernation by warm sunny days, a bumble bee forages in the early-flowering *Crocus tommasinianus* 'Ruby Giant'. Early-flowering geophytes such as snowdrops, *Crocus* sp. and *Scilla sibirica* suit T-lawns very well and provide much-needed energy for pollinators early in the year.

It is possible to include great swathes of geophytes in T-lawns since mowing rarely starts before the spring equinox, allowing for early bulbs to largely complete their growth cycle; their leaves having mostly withered or lost their efficacy by the time the mower collects them.

Later in the year, especially after mid-summer, it tends to be the non-native plant species that offer most floral resources to pollinators, since most British natives have passed their floral peak. Plants such as naturalised *Pilosella aurantiaca*, southern Hemisphere *Lobelia* sp. and Himalayan or African *Parochetus communis* continue the supply of pollen and nectar well into the autumn when native-borne supplies are dwindling and becoming scarce. This is a good argument for the inclusion of non-native plant species in T-lawns, since without them, pollinator resources in the lawn diminish in late summer as native flowers move toward producing seed and storing carbohydrates for the coming winter.

5 How to Make One

The ideal approach is for tapestry lawns to be started as a blank canvas, using an area that has been cleared of any previous botanical occupants and preferably with a seed tray depth layer of the fertile topsoil removed too.

Since traditional lawn grasses like to have well-fertilised soil, it might seem a bit odd to remove what is often considered the best bit of soil in a garden to lay a new lawn (although it doesn't always have to be removed if you already have a particularly poor or thin soil). However, topsoil is usually the most nutrient rich part of a soil profile, it will tend to be peppered with the remnants of its previous occupants in the form of rhizomes and tap roots and it is very likely to have a well-stocked seed bank too. Removing this layer can reduce the likelihood of the reappearance of any of the previous forbs and grasses that inhabited the spot and thus reduce any lawn weeding you may need to engage in at a later date.

SOIL FERTILITY AND SUPPLEMENTS

Just as with the plant communities of the Scottish machair, a reduced or low level of fertility is helpful in maintaining diversity in T-lawns. This may seem strange if you've always listened attentively to the advice of gardening gurus and worked toward having a fertile soil by adding well-rotted manure or garden compost and the like, or you have previously fed your grass lawn with synthetic fertilisers to keep it looking healthy and lush. But with a tapestry lawn you will now be aware that we aren't looking to cosset the plants and encourage lush growth. It will only mean there's more mowing to do, and a high level of fertility will really only suit those plants that do well in fertile sites; buttercups will love it!

Since we are creating a species-rich plant community rather than encouraging each individual plant in a garden to grow and bloom to its maximum potential, we ideally need a soil that isn't nutrient rich but also isn't entirely devoid of nutrition – something of a goldilocks soil for T-lawns. If it is completely devoid of nutrients, no useful plants are likely to thrive, and if it is your average garden-type fertile soil it may potentially be too rich and favour nutrient-hungry plants whose nature is to process nutrients fast, competitively grow fast and overwhelm slower growing smaller species that don't need or are not able to utilise high levels of nutrients.

It may seem counterintuitive since we are so often told to fertilise plants for best performance, but a T-lawn on very fertile soil or one that subsequently receives fertilisers will likely be a leafier and less floriferous lawn than one that is on poor soil and that remains unfertilised. T-lawns do not need any form of supplementary fertilisation; it upsets the community balance, and it should not be undertaken. Mowing requirements are also increased with fertile soils since well-fed plants tend to grow faster, and this inevitably will mean mowing off a lot more flower heads more frequently. This may therefore mean that some species do not get the chance to produce a floral display since you should be mowing when competition becomes evident.

You are most likely to find slightly poorer soil in your garden by removing around the depth of a seed tray – about 3 cm (1¼″) of topsoil, probably more if you've been adding garden compost for years and years in your chosen spot – in which case you might also think of additionally turning over the ground with the aim of bringing some of the less nutrient-rich deeper subsoil to the surface.

It is a good idea to discover what your soil profile is like with an auger if you have access to one. Builders can cover things up with grass turf, and there can be all kinds of potential problems hidden underground that range from pure sand and builder's rubble to layers of clay, rocks, pipes, and chemical residues. Knowing about the subsoil in advance can help determine what needs to be done, if anything, and what plants will likely suit the site. A compacted soil underneath laid plant tiles can restrict plant development considerably (Figure 5.1) and should be amended prior to laying the lawn.

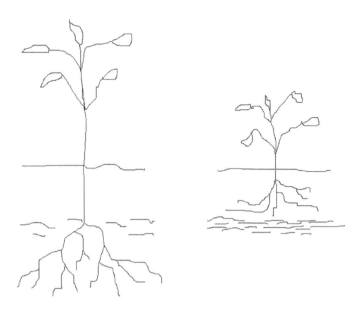

FIGURE 5.1   Plant tile trays root best into soil that is not compacted.

If you don't fancy shallow digging the spot by hand, a rotovator can aerate compacted soil rather well and can mix up the remaining topsoil and subsoil to create a sort of in-between-type blended soil (Figures 5.2 and 5.3). No matter what your soil type, aerating the ground prior to laying the lawn is good practice. The more times you go over the spot with the rotovator, the more evenly the soil will be mixed; however, some variation across the site is not a bad thing.

FIGURE 5.2   The previously grassed-over turning roundabout at Myerscough College being prepared by project student Chris Piner using a rotovator to loosen the soil. Later the soil was raked to level the ground prior to laying the plant tiles. The grass turf was previously lifted and removed and, in the process, revealed a localised layer of clay that required breaking up before the lawn could be laid.

FIGURE 5.3 Prepared ground at Avondale Park, Kensington & Chelsea, London. The ground had been cleared of the previous wildflower meadow, and although the topsoil was not removed, it was rotovated twice over and raked. The soil is an urban clay loam, and there is a gentle slope from the back wall towards the path.

It is at this point you may consider adding and blending in any of the often-recommended physical drainage improvers, such as horticultural grit and coarse sand, if the site is typically prone to waterlogging. However, poor drainage usually means either that your water table is high or, as is often the case with clay soils, that the water has nowhere to drain to. Adding grit and sand to a poorly drained spot is rarely effective; huge amounts are needed. The Royal Horticultural Society (RHS) recommend 250 kg (550 lbs) per m^2, and it tends to create a sump that water from surrounding impervious soil accumulates in.

A simple test for soil water issues is to dig a hole 60 cm (2 ft) deep. Cover it and leave overnight. If there is water in the bottom in the morning, you have a high water table. If it is dry, fill with water and observe if it drains away overnight. If water remains in the hole, this indicates poor drainage and suggests a clay soil and/or compaction.

There is the option to use a raised-bed type approach such as that used for the exhibit at RHS Chelsea in 2013 (Figure 5.4), whereby the entire area of the lawn is elevated above ground level and suitable low-nutrient loam-based substrate used. Drainage can be improved immediately by this method.

If needs be you might consider implementing something of a slope on which to lay your lawn with appropriate drainage at the base of the slope to take the water away. The original inspiration for tapestry lawns came from a poorly drained, summer-cracking Oxford clay lawn on a slope with a small and notoriously inefficient driveway drain nearby, or you might just risk it and be a little more selective in your choice of plants, choosing first from those you can see doing well in your local area and then adding those that should have a fighting chance in wetter soils.

Removing the top 3 cm (1¼″) or so of soil can usefully lower the soil level by about the depth of a seed tray and will be topped up again to its original level when the lawn is laid. If you've had to rotovate and the soil has fluffed up a bit, this may not be the case. The laid tapestry lawn may be proud of the original soil level but will usually sink back close to its original ground level as the lawn settles (Figures 5.3 and 5.4).

FIGURE 5.4 Demonstration T-lawns in raised beds (shown on pallets) being grown for the 100th RHS Chelsea Flower Show 2013.

Levelling the soil surface as best you can before laying is a good idea since a lawn mower will be used to closely manage the finished article, and notable lumps and bumps may cause the lawn to be scalped by the mower's blades. This can be done by raking so the surface is broadly even. It is helpful if the surface is also relatively loose and left scarified by the rake to allow easy penetration by the roots of the plants laid on top.

Previous lawns have used readily available peat-based multi-purpose garden substrates and low-nutrient seed mixes to grow the plants. A peat-lite seed substrate containing 50% fine vermiculite and based upon the Cornell University mixes for commercial growing has proved particularly useful [48]. Peat-free substrates have not been specifically used for T-lawn propagation due to the wide variability in such products during the initial research into the format. Consistency in the materials used during scientific research is essential if results are to be comparable and replicable in any meaningful way. With the recent advent of some more reliable and consistent peat-free substrates, it now seems an option to propagate using them. In another horticultural oddity, both peat-based and peat-free growing mixes are usually called composts, although the method of manufacture is variable and if peat based they are not actually completely composted. True garden compost is usually too variable in composition, structure and nutrients to be used.

Experience suggests that substrates used specifically to start seeds are particularly useful; they are low in nutrients and these are soon used up, and at planting, plant roots move quickly to obtain nutrients from the native soil, but it is not the perfect substrate. Peat-based substrates can be subject to drying surprisingly quickly, and newly laid lawns will need watering until they are established. The use of peat is also rather frowned upon these days. Observations over several years indicate that the covering layer of organic substrate provided by the tiles will become incorporated into the native soil within a couple of years by the activity of soil organisms, especially worms, so long as it does not dry out. It seems reasonable to suppose that if worms are in short supply that soil incorporation may take longer.

The tiles also have another useful function in that they act to put a cap on any seedbank held within the native soil. Any weed seeds in the soil will be buried under the depth of the plant tile. Those weed species that require light to germinate are unlikely to show themselves later on, and even the occurrence of perennial weeds is greatly reduced, unless they happen to be re-sprouts from pieces of tap root left in the native soil. It is worth spending a little time carefully extracting perennial weeds from the chosen spot before undertaking to lay the new T-lawn and diligently picking out any perennial tap roots that might be seen after rotovating and during raking.

Some plants don't readily make a strong root mat in the trays, for example, *Acaena* sp. and some *Viola* sp., and at planting the substrate may crumble away from roots as the plants and substrate are removed from the trays for laying. A useful double-flip methodology was spontaneously introduced in Dorchester, UK, at a community T-lawn planting day by a T-lawn enthusiast who found that for crumbly substrate with poorly rooted plants that upending the tray onto a flat surface such as a small tray-sized board, for example, a handled seed tray press, she could then with a quick wrist action flip-lay the tile without too much disturbance (Figures 5.5 through 5.9).

FIGURE 5.5   To assist laying of plant tiles a board can be laid on the top of the plant tile.

FIGURE 5.6   The plant tile is upturned onto the board.

FIGURE 5.7   Brought adjacent to the site intended for the plant tile.

FIGURE 5.8   The plant tile is dropped into place.

FIGURE 5.9   Planted tile.

The plants can become squashed having been upturned on the board, but these are plants that will be mown and walked upon in the future, so they can certainly take being squashed upside down for a moment or two, especially since after the lawn has been laid it is well worth rolling it with a heavy garden roller to even it out (Figure 5.10).

Rolling is an activity unique to lawns and has two useful effects. As in the preparation of traditional grass lawns, it acts to level the lawn surface, and additionally with T-lawns the pressure placed on the tiles makes them spread sideways, consolidating the join between them and filling any gaps that might occur. This is a relevant and useful activity since poorly linked tiles can act as a barrier to root and rhizome penetration, and any gap can act to dry the edges of tiles. If a roller is not available, then walking in short steps on the tiles has the same effect although does not provide as evenly distributed pressure.

Once the lawn has been laid and rolled it can look just a little messy since it hasn't yet had time to knit together for full ground cover effect and will be a mix of very full and sometimes overflowing tiles and some less so, all with roller-squashed and foot-squashed plants.

FIGURE 5.10   Rolling newly laid plant tiles can assist in making a good connection between the tiles and the native soil and acts to slightly squash and spread the tiles, ensuring a good connection between them. This is best done when tiles are moist. Here rolling is being applied to a newly laid extension of an existing tapestry lawn.

Squashing, along with occasional footfall on tapestry lawn plants, is not a bad thing since it has the effect of pressing the plants into the surface of the soil encouraging adventitious rooting at nodes, and it can affect the plants' subsequent architecture. The occasionally well-placed boot on initially protruding plants has proved effective in levelling uneven tiles and any single pot plants that may be proud of the surface, although it is certainly not what most people expect as a planting methodology. The plants will recover and the tiles will blend together over the next 6 weeks or so as they become established and root into the native soil (Figures 5.11 through 5.15). It is particularly important during this establishment period that the lawn is not allowed to dry out completely, or establishment can take substantially longer and some of the shallow-rooted species may suffer losses.

FIGURE 5.11   Trays of plants being laid at Avondale Park, London.

FIGURE 5.12   Once laid, the Avondale Park T-lawn receives a much-needed watering.

FIGURE 5.13   A moveable sprinkler was used subsequently over the next 6 weeks. Tiles on the dry side weigh less and are therefore easier to transport. Rolling after watering the new lawn is more effective than rolling immediately after dry-laying, and it is an effective method of ensuring that the tiles fit together since wet tiles squish together much easier than dry ones, a bit of boot-on-plant tile action can be helpful too.

FIGURE 5.14   The Avondale Park lawn once established.

FIGURE 5.15   A slightly closer view of the Avondale lawn. The variety of leaf colours and textures are just as important as the floral display.

Occasionally some species may respond poorly to being moved from where they were grown to the new conditions that exist where they are laid; this is generally referred to as transplantation shock. *Parochetus communis* (blue pea/shamrock) and *Micromeria douglasii* (Indian mint) have been observed to lose their leaves entirely, most especially if planted in an unsuitable spot, but only rarely do the plants immediately die; they generally re-sprout with a new set of leaves appropriate to their new situation. If grown in situ, this disturbance effect does not seem to occur.

If plant tiles are unavailable or impractical, it is possible to use the technique of cluster planting with similar results. This method requires that several pot-grown plants (or dense groups of large plugs) are planted together in close groups. Groups of three and four pots have been used with satisfactory results, and in very small T-lawns, single well-developed pots can be used (Figure 5.24), although the proportions of the plants used becomes particularly important, as competition swiftly changes the layout of the lawn. One buttercup to one daisy is asking for trouble!

GARDENERS WILL PLAY

Usually lots of plant tiles are needed for a classic-style T-lawn to cover a good-sized lawn, but smaller lawns can be achieved with a bit of determination and the ingenuity frequently found in enthusiastic and creative gardeners (Figures 5.16 through 5.26).

Lawns can use a mix of techniques and can be made to suit the outcomes of their creator.

On an overall larger scale but using several small lawns, combinations of all methodologies have been used to good effect, although a bit of time is required before the lawn offers complete coverage.

FIGURE 5.16   A small front garden lawn grown in situ in trays and newly laid. (Courtesy of L. Tattersall, 2017.)

FIGURE 5.17   A default grass lawn is about to get the T-lawn treatment. (Courtesy of K. Laing, 2015.)

FIGURE 5.18   True determination. Trays of plants are grown, and a combination of pots of good-sized plants and multiple plugs used as well. A stash of additional trays is also growing in a small plastic zip-up greenhouse out of sight. (Courtesy of K. Laing, 2015.)

FIGURE 5.19   The finished article. A mixed environment to suit the owner's requirements. (Courtesy of K. Laing, 2015.)

FIGURE 5.20   A mix of cluster planting and tile planting used at Oakham School. (Courtesy of R. Dexter, 2015.)

FIGURE 5.21   The use of cluster planting and tiles together can be clearly seen as this new lawn begins to knit together. (Courtesy of R. Dexter, 2015.)

FIGURE 5.22 Four small L-shaped T-lawns are incorporated around a central space in a commissioned design at Oakham School. The head groundsman notes it has been much easier to maintain than was expected, is used by teachers as part of their science classes and is much commented on by visitors. (Courtesy of R. Dexter, 2015.)

FIGURE 5.23 The tapestry lawn installed at RHS Wisley in May 2018. A combination of 14 plant species in a unique pattern contained within a traditional grass lawn. The plants selected by RHS Wisley were *Leptinella squallida*, *Lobelia pedunculata* 'County Park', *Acaena inermis* 'Purpurea', *Thymus serpyllum*, *Lobelia pedunculata*, *Mazus reptans*, *Lysimachia nummularia* 'Gold', *Lotus corniculatus*, *Achillea millefolium* 'Summer Shades', *Phyla nodiflora*, *Viola hederacea*, *Parochetus communis*, *Potentilla reptans* 'Pleniflora', *Argentina anserina* 'Golden Treasure' and *Ranunculus repens* 'Gloria Spale'.

AN ALTERNATIVE METHOD?

Private lawnscape managers (gardeners) keen to have their own T-lawns are not always able to produce plant tiles or obtain large numbers of suitable plants for chequerboard or cluster planting. Being an adventurous lot, other methodologies have been used, and although not specifically rec-ommended here since such methodologies did not always produce suitable results during the initial experimental work, they have been employed with some relevant results. Scatter-planting where single pot plants are widely spaced has been one such methodology (Figures 5.24 through 5.26).

Scatter-planting does not provide instant coverage and requires time for the plants to spread into each other. It also means that the plants are not generally mown for some time and develop their usual phenotype forms rather than the mowing responsive form of fuller initial coverage methods. It is a method that can clearly demonstrate the vigour of the plants used. In the initial stages, it is possible to see which plants are spreading faster in the local conditions and allow for responsive thinning out of the very vigorous plants and the addition of more of the plants that are slow to spread and likely to be overwhelmed.

Mowing a lawn created in this way for the first time can result in notable damage to the individ-ual plants and in some cases cause plant death. Since essentially the plants will have developed as individuals rather than as a member of a tight-knit community, the effect of mowing can be surpris-ingly drastic, especially on the taller growing plants. This tends to be due to a behavioural oversight on the part of the lawn's keeper since the application of mowing will likely have been delayed by referring to the growth and floral performance of individual plants and not fully considering the yet-to-form community.

FIGURE 5.24   Scatter-planting a large area is a technique implemented by T-lawn enthusiasts with limited budgets. (Courtesy of G. Baldwin, 2014.)

FIGURE 5.25   The use of well-developed plants in single pots spread over a designated lawn space is a method that can produce interesting results. With insufficient plants to initially cover the area, the plants must spread and join up to create a lawn, and the space between plants must be constantly monitored for weeds while this happens. Since time is required for the plants to meet up and blend, their architecture can be drastically affected when first mown. (Courtesy of G. Baldwin, 2014.)

FIGURE 5.26   After a year the scatter-planted lawn has developed its own character and has the look of a more conventionally planted T-lawn. (Courtesy of G. Baldwin, 2015.)

The key triggers that indicate the whole lawn needs mowing and that are made evident by the behaviour of the whole community are largely missing. It may be possible to avoid the worst by having the steely resolve to mow the entire cadre of scattered pre-lawn individuals and continuing to do so before and during the period that the plants meet up and finally knit together to become a lawn. In this way, the plants may essentially be preconditioned.

SUMMARY

- Ideally start with a space bare of grasses and other plants.
- Poor native soil is not a bad thing but compacted soil is.
- Remove the top 3 cm of fertile topsoil if appropriate.
- Remove any perennial weeds carefully.
- Dig over or rotovate the remaining soil to improve aeration, and rake it level before laying to allow for root development.
- Continue to remove perennial weed roots while raking.
- Lay the plant tiles you have selected for your lawn closely together.
- Pay attention to the immediate environmental conditions when choosing where to lay plant tiles; a sun-loving plant will not do well in a generally shady spot and vice versa.
- Plant tiles can act as a useful cap to the seedbank in the native soil.
- Lawns should be wet-rolled after laying to both level the surface and to ensure a good join between plant tiles.
- Keep the newly laid lawn well-watered for the first 6 weeks while the plants establish.
- Do not use supplementary fertiliser on your lawn.
- Cluster planting with groups of the same species or cultivar can be an alternative method but requires diligent removal of unwanted species and additional time before full coverage is achieved.
- Ingenuity and determination are valuable items in a lawn-keeper's arsenal.

6 Maintenance

If there is no ongoing maintenance of a tapestry lawn, there will soon be no tapestry lawn to speak of. Along with initial poor plant choices and proportions, this is by far the most frequent reason for T-lawns to degrade or fail. Sadly, after several years of general neglect and not being weeded or mown appropriately, the originally much-admired T-lawn in Avondale Park, London, has ceased to be either a tapestry of chosen plants or a lawn.

There are four significant management areas essential for the ongoing maintenance of a tapestry lawn:

- The continuous removal of grasses (and other weeds)
- Mowing
- Maintaining the proportionality of suitable species
- Watering

REMOVAL OF GRASSES

A richly planted and densely matted T-lawn can be surprisingly resistant to invasion by grasses, but it is not immune by any means. In just the same way that dandelions (*Taraxacum* sp.), clovers and daisies turn up in a grass lawn as mowing-tolerant weeds, so mowing tolerant grasses becomes the primary weeds that can turn up in a tapestry lawn. *Poa annua* (annual meadow grass) is a particular bugbear along with *Elymus repens* (couch grass). Grass seeds will be carried to your T-lawn on the wind, by splashes of water, on shoes and wheels, by animals and birds and all manner of other surprising vectors. Apart from fully enclosing your lawn in a protective bubble, there are two common methodologies that can be used to address weeds in the lawn, and these are used in both grass and tapestry lawns.

REMOVAL BY HAND

This is both effective and with patience not particularly difficult to do, especially for smaller lawns. The different architecture between grasses and forbs makes grasses easy to identify, and the best approach is to simply pull or tease out the grasses from the lawn whenever you see them. A small gardening knife is good for this sort of activity, and slicing just under the grass's meristems rather than necessarily digging them out completely has been effective (Figures 6.1 and 6.2). The idea being that you remove them before they can set seed or vegetatively spread, particularly *Poa annua*, which can set new seeds every 10 days or so, in as little as two months from it first appearing. This weed life cycle sets the general tone for how often weeding at a bare minimum level should be undertaken, which is at least once every couple of months. Ideally it will be done whenever a weed is spotted. This is an ongoing process but can be particularly effective in spring (and in late autumn) when the tall slender blades of grass will often poke up above the forbs and be easily spotted.

HERBICIDE

Just as there are broadleaf lawn weed killers (selective herbicides) that target non-grasses in grass lawns, 2,4-D being the original example, there are selective herbicides that target only grasses. For a variety of reasons, at the time of writing this, these can be difficult to come by in the EU; however Laser®, a graminicide produced by BASF, has in 2017 been licensed for use in ornamental horticulture and may be used appropriately by qualified professionals. It is not available for home use. The situation is different in North America where ornamental horticulture has access to several additional different

FIGURE 6.1 Rye grass (*Lolium perenne*) growing amid a clover-flowered T-lawn. The architecture of grasses makes them easy to identify within T-lawns. Relatively large species can be tugged out by hand, and smaller finer species removed by using a suitable small blade applied just below the ground level meristems. It is frequently much easier to 'slice out' the finer grasses than to attempt to dig them out.

FIGURE 6.2 Once the grass has been removed, the forbs in the lawn will swiftly fill the gap.

products such as Acclaim®, Fusilade® II, Vantage® and Envoy®. All graminicides require mixing with an adjuvant, something that makes the active chemical adhere to the plant and not simply run off.

Like most herbicides, graminicides are not always 100% effective since some species can be resistant to the active ingredient (a surprisingly large number of grasses are), and dosage can be variable depending on the grasses you are trying to control. As ever, always follow the manufacturer's instructions. It is always best to check over your lawn a week or so after application for any grasses that have escaped their chemical doom.

MOWING

Mowing is not only necessary for ecological purposes in its ability to facilitate the coexistence of plants with different habits, but aesthetic ones too. A lawn has a low aspect, and although tapestry lawns cannot be expected to maintain a levelled playing field look, they do require mowing in response to increasing height.

Mowing can take place at any time between the spring and autumn equinoxes; it is not usually required much before or after those dates in the UK. Before the spring equinox, the lawn may be richly enamelled by early flowering geophytes such as snowdrops, early crocus, *Scilla* and early miniature *Narcissus*. These plants have bulbs and corms that require rejuvenating after flowering and require that ideally their leaves remain intact until they fade away of their own accord. Inevitably the environmental conditions will have an influence on how long the leaves remain and how successful they have been at energy production, but most of the early flowering geophytes will have managed to store sufficient carbohydrate by the start of May in the northern hemisphere for reblooming the following year and may be mown even if they are still showing some green leaf without significantly harming the plants. Early-flowering *Crocus* sp. have been repeatedly observed to set seed and produce seedlings in T-lawns, with populations increasing over time.

The frequency of mowing will be down to the plants that have been incorporated into the lawn and the environmental conditions, particularly altitude, soil fertility and the availability of water, but mowing can be expected to be undertaken around three to five times a year. In 2012, a year noted for its dismally regular and heavy rainfall (even for the British Isles), trial lawns had to be mown two times more than usual to manage the particularly lush growth. This lush growth highlighted the need for the ongoing maintenance of sharp blades in mowers used to cut T-lawns, since lush stems and leaves can become difficult for dull blades to cut efficiently and cleanly.

The choice of mower is not restricted; rotary and cylinder mowers that can collect arisings both work well, although the best results have been achieved by powered mowers rather than hand-pushed ones. The cleanest cuts and even grass-lawn-type parallel lines can be achieved by cylinder mowers, but the lines are not as clear and crisp, or as long lasting as those created on grass. Arisings must be collected by the mower and removed from the lawn. Arisings are leafy and tend to be high in nitrogen in a similar manner to grasses and should be treated as such when composting. Raking after mowing is not an option since raking tears and tugs at the mingled plants and effectively acts to destroy the interwoven community. This is something to bear in mind if your lawn should be close to deciduous trees; raking the lawn of leaves in the autumn is not an option. If leaves are problematic, a soft broom or leaf-blower can be used for the job or you might just leave them to the worms if you can bear to lose sight of your lawn for a while.

Ideally the height of the mower should be set to around 4 cm to both maintain a lawn-like appearance and facilitate plant coexistence and survival. Mowing should be initiated when competition is becoming clearly evident and low-growing plants are clearly being shaded and overwhelmed. Sometimes this will mean that a sea of flowers will be dramatically decapitated, but this process is essential to the long-term survival of the lawn. It is a lawn and not a flower bed. You must mow one sea of flowers to get another, albeit a different one. Delay in mowing will likely cause damage

FIGURE 6.3   With the lawn keeper beguiled by an attractive floral display, this T-lawn was left unmown for too long. Once mown, the result is an untidy and unattractive show of etiolated and lightly woody stems. It is likely that the lawn will recover but the situation could have been avoided by earlier mowing.

to the internal structure of the lawn and can result in poor aesthetics both in the short and longer term (Figure 6.3). Once mown, if the lawn looks like a multi-tonal lawn and has good coverage and a pleasing neat aspect to it, then you are probably mowing it about right. Tidy edges as necessary.

Without mowing, the lawn will change considerably. The taller-growing plants will lift the height of the lawn, genotypic plasticity will see some species change their growth habit as they compete for light and space and many of the prostrate plants will die out. In effect, it will become meadow-like. Unplanted species, especially those with wind-distributed seed, such as trees, nettles and thistles, may appear now they are no longer environmentally filtered and prevented from establishing by the cut of the blade (Figure 6.4).

FIGURE 6.4   A year or more of neglect and a tapestry lawn becomes unrecognisable. Prostrate plants have largely disappeared, wind-distributed plants have taken root and the growth of the remaining plants is substantially taller. This is not a situation that is easily reversible.

MAINTAINING SUITABLE SPECIES PROPORTIONALITY

Each lawn will have a unique set of environmental conditions. Inevitably some of the plant species used will find it congenial and do very well, sometimes very well indeed, and to the detriment of some of the other species. For example, a run of several very wet years, a particularly fertile soil or poor drainage might lead to creeping buttercups spreading more than we might like, and any species with niche adaptations for drier, nutrient-poor or well-drained conditions such as creeping thyme or silver-leaved (*niveum*) *Pilosella tardans* may well become extinct. In those kinds of circumstances and bearing in mind that a species-rich lawn is generally a better-looking and better-behaved lawn, a bit of lawn gardening is most probably required.

WATERING

Watering requirements are largely dependent on the local weather, soil conditions and the plants included in the lawn. Most established T-lawns rarely require watering except in unusually dry periods and will usually continue to remain green as the grass lawns around them go brown in mid-summer. However, new lawns should not be allowed to dry out during the period of establishment, which is roughly 6 weeks or so.

Unlike shallow-rooted lawn grasses, which will naturally brown and die back to ground-hugging dormant meristems during dry periods and then re-sprout when conditions improve, most lawn forbs can be damaged and killed by drought. Amongst many species-specific adaptations, the relatively deeper rooting capacity of many lawn forbs aids the plants in staying greener for longer than is generally seen in lawn grasses, and some have the capacity to lose their foliage and die back to tap roots and rhizomes, but many will simply dehydrate beyond their capacity to recover if their environmental tolerances are exceeded. As a general rule, if during the day the three leaflets of *Trifolium repens* in the lawn are deflated like a collapsed umbrella, it is almost certainly time to get out the sprinkler. Waiting longer, for example, for the leaves to go crispy, is likely to reduce the recovery of the lawn, and plants may be lost.

LAWN GARDENING

There is no specific requirement to wait until a species has become problematic to tweak your lawn. Tweaks can be made whenever you as the lawn keeper feel it necessary to suit your personal aesthetic.

Inevitably in a mixing and mingling plant community the initial distinct patches will merge together, and the clear distinct patches of a new lawn can blend into something of a more homogenous-looking lawn (Figure 6.5).

This doesn't always happen. Areas of a lawn can develop distinct character and maintain them for some time, but after 2 or 3 years, a very mixed community is likely to have formed in some parts of your lawn. This is not in any way a bad thing. The lawn will have self-organised, and plants suitable to your environmental conditions will have become apparent in areas that suit them. These will guide you in future plant choices and give a more accurate idea of your lawns conditions than perhaps you might have originally imagined.

It may, however, have lost the 'wow' factor it may have started with if visual aesthetics are the most important to you. If you are simply happy to be the proud owner of an eco-friendly, species-rich lawn and not too bothered about the overall mixed look of the lawn, then simply continuing to mow, weed and maintain species proportionality should see your lawn lasting quite some time. However, if you wish to refresh the lawn, then there is a simple technique that will return patches and allow you to resolve other visual issues.

FIGURE 6.5 A three-year-old tapestry lawn showing how the mixing and blending of plants can lead to what can appear to be a more homogenous appearance than the original planting, especially late in the year. The lawn here retains over 30 different species, but the initial patches have spread and a more traditional lawn-like appearance is evident.

It is best undertaken during the growing season as early as possible to allow your new plants to establish or when the mood for change or the current aesthetic prompts you. Although it is worth remembering that most British native plants tend to become apparent in waves in T-lawns, putting on most of their growth as they move toward their flowering season and then seemingly halting and even dwindling once that season has passed. What can seem like overwhelming coverage by one species during its floral period can a month later (or after being mown) seem much less so, as it returns to a more vegetative state and moves from a floral contribution to a ground cover contribution. Your supposedly troublesome lawn plant may not be overly dominant but rather be a display of a plant in its puffed-up, notice-me floral season. The same can apply to the raggedy or spent look that some plants get after flowering and being mown; the plants can still be fine, even if they look a bit past their prime, which of course they are.

The process is a simple one but does require some preparation in the form of growing, unless you happen to be able to get hold of plant tiles (or pots) of the species type you wish to add. Simply use a spade to cut into the lawn to the depth and shape of a plant tile, work the spade's blade under it to loosen it and then lift it away. Replace what you have removed with the new plant tile, firm it in with your dainty boot making sure the edges are in close contact with the lawn and water well (Figures 6.6 through 6.11).

Continue to water the new tile as necessary until you can clearly see signs of new growth, since this is an indication that it is obtaining sustenance from within the lawn itself. Spring and early autumn are ideal times to both lay new lawns and to make changes to existing lawns, as the soil is warm enough for new roots to develop, and climatic stresses are generally at their most moderate.

FIGURE 6.6   Here an area of poor coverage needs some attention.

FIGURE 6.7   Use a spade or other suitable tool to cut out an area equivalent to the size of a plant tile. This methodology is also relevant in an area that has become dominated by one species or for the simple addition of a new species or cultivar. It can be used to refresh a lawn that has become more homogenous in appearance than is desired.

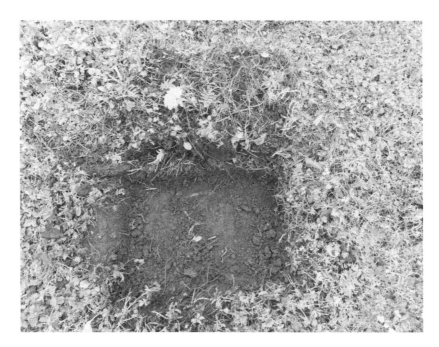

FIGURE 6.8   Remove a tile-sized area to the depth of a seed tray. It is at this point that you may notice how the original substrate layer provided by the initial trays has all but disappeared, having been incorporated into the soil by the action of soil-dwelling organisms such as worms.

FIGURE 6.9   Fill the space with the replacement plant tile and press firmly into place.

FIGURE 6.10   Once in place, the new plants should be watered and ideally allowed to establish before being mown.

FIGURE 6.11   A week or two later, the new tray is blending well with the lawn.

7 Outcomes

Each lawn will develop and self-organise in an unique fashion based primarily on the local conditions and the plant choices made at inception. It is therefore difficult to foresee what any one lawn will look like as it develops over time. It is generally the case that most plant tiles survive an initial full annual set of seasons, and the lawn will look quite striking in its first year as the individual species and cultivars are clearly distinct (Figure 7.1). As the lawn self-organises and plants move and mingle appropriate to their niches and the conditions in which they find themselves, the initial layout will blur and restructure itself. Plants unsuited to the initial spot are likely to move via vegetative creep and occasionally by seed, or they will die. During this process, some eye-catching and interesting outcomes might be achieved (Figures 7.2 through 7.26).

FIGURE 7.1   An establishing T-lawn prior to its first mowing.

FIGURE 7.2   From a distance, the complexity of a T-lawn is not often fully appreciated.

FIGURE 7.3   Move in a little closer and much more of the complexity is revealed. In this 3-year-old T-lawn, a dozen species mix and provide complete ground cover.

FIGURE 7.4   In another part of the same 3-year-old lawn, the *Ranunculus repens* cultivar 'Gloria Spale' provides an alternative to the usual bright and reflective yellow typical of the species.

FIGURE 7.5   A very simple combination of *Bellis perennis* cultivars and the typical bright and reflective yellow of *Ranunculus repens* can provide an unexpectedly pretty combination.

FIGURE 7.6 When viewed in close-up, a T-lawn can be particularly interesting. Here the mixed leaves of *Leptinella squallida* 'Platt's Black', *Viola hederacea*, *Ajuga reptans* 'Burgundy Glow', *Lobelia oligophylla* and *Chamaemelum nobile* are the primary players.

FIGURE 7.7 Mostly ornamental *Trifolium repens* with a splash of gold from *Ranunculus repens* 'Buttered Popcorn' and the pale-blue stars of *Lobelia pedunculata* flowers.

FIGURE 7.8 A mix of the microclover *Trifolium repens* 'Pipolina', *Dianthus deltoides* 'Flashing light', *Leptinella dioica*, *Lysimachia nummularia* 'Aurea', *Prunella vulgaris* and *Chamaemelum nobile*.

FIGURE 7.9 The use of cultivars with coloured foliage and clear leaf forms can be used to good effect. *Lysimachia nummularia* 'Aurea' and *Acaena inermis* 'Purpurea' mix with the green-star leaves of *Potentilla reptans* 'Pleniflora'.

FIGURE 7.10 Mostly *R. repens* 'Buttered Popcorn' and *Lysimachia nummularia* 'Aurea' creating a patch of gold in the lawn.

FIGURE 7.11 With most T-lawn plants having a creeping habit, eyesores such as this drain cover can be covered surprisingly quickly. Creeping microclover *T. repens* 'Pipolina' and *Stellaria graminea* are doing most of the work here.

FIGURE 7.12   A colourful mix of foliage types and a few *Trifolium repens* flowers.

FIGURE 7.13   A contemplative T-lawn, ready for the mower but looking attractive.

FIGURE 7.14 In a sunny spot, a mixture including *Geranium pyrenaicum* 'Bill Wallis', *Argentina anserina* 'Golden Treasure' and *Chamaemelum nobile* 'Flore Pleno' does particularly well. Indigo is a rare colour in T-lawns, and *G. pyrenaicum* 'Bill Wallis' can be a relatively tall lawn geranium, but it seems to behave itself when sparsely scattered, planted individually or in small groups and mown.

FIGURE 7.15 In early spring, *Primula veris* and *Primula × pruhonicensis* 'Wanda' flower alongside the wild form of *Bellis perennis* and its cultivar series 'Pomponette'. *Ajuga reptans* 'Atropurpurea' is also flowering. Primulas have generally finished flowering by the time the first mowing is applied, and losing a few older leaves to the mower does not appear to be detrimental to them appearing again the following year. Since setting seed is usually rare for *Primula* sp. in a T-lawn, they benefit from being topped-up from time to time.

FIGURE 7.16   Spring can bring a riot of colour. It is when the *Bellis perennis* cultivars tend to be at their best, as they use the carbohydrate stored over winter in their roots to flower their hearts out. Here red 'Pomponette' cultivars steal the show. What might appear to be grasses are the leafy remains of early-flowering *Crocus* sp. Some cut narcissus leaves are also visible, but these were a mislabelled large-flowered variety and a somewhat later-flowering variety than is appropriate and were later removed.

FIGURE 7.17   Spring, with *Bellis perennis*, *Viola odorata* and a red-leaved *Trifolium repens*.

FIGURE 7.18 In the Avondale Park lawn shortly after its inception, the use of plant tiles provides clear patches and can give the impression of drifts, showing the value of both coloured foliage and flowers.

FIGURE 7.19 A closer view showing *Trifolium repens* 'William' (Top), and 'Wheatfen' (middle) intermixed with *Veronica chamaedrys*. Red-flowered *T. repens* 'William' can be surprisingly vigorous in ordinary soil and frequently becomes a trigger for mowing due to its tendency to produce long leaf petioles (stalks). It is somewhat better behaved in poorer soils, and 'Wheatfen' tends to be much better behaved overall although it has typical white flowers.

FIGURE 7.20 Late summer when the main floral periods for most British natives have passed and overall growth in the lawn has slowed. Suppressed by mowing earlier in the year, here naturalised European native *Pilosella aurantiaca* has been allowed to grow taller than its companion vegetation, and the *Achillea millefolium* is also growing taller. This allows for its late, orange flowers to appear in the lawn, and in the process the circumstances have facilitated an errant *Plantago media* (hoary plantain) that has been allowed to grow. The flowers of *Lobelia pedunculata*, *Veronica chamaedrys*, *Potentilla reptans*, *Geranium pyrenaicum*, *Phuopsis stylosa*, *Stellaria graminea* and a pigeon-munched *Trifolium repens* 'William' provide the floral background.

FIGURE 7.21 In the same lawn, located in front of a collection of bee hives, the late summer blooms of *Pilosella aurantiaca* and *Chamaemelum nobile* 'Flore Pleno' will be shortly followed by *Achillea millefolium* and offer an additional resource to late-foraging honey bees. It is a bit naughty to allow the taller growth, and there may well be ornamental consequences; however, this lawn has a particular bee-friendly purpose and is not strictly ornamental.

FIGURE 7.22   A summertime mixing of a red-leaved and flowered *Trifolium repens*, *Argentina anserina* 'Golden Treasure', *Chamaemelum nobile*, *Ranunculus repens* 'Buttered Popcorn', *Dianthus deltoides* 'Arctic Fire' and some mysteriously appearing, non-flowering and rather unhappy *Dianthus gratianopolitanus*.

FIGURE 7.23   In a sunny, dry spot plants with similar niche tolerances have self-organised and mixed together. *Thymus praecox*, *Acaena microphylla*, *Leptinella potentillina*, *Pilosella tardans*, *Argentina anserina*, *Geranium thunbergii*, *Potentilla reptans* and *Pilosella officinarum* are the prime species.

FIGURE 7.24   A recently mown lawn can still contain blooms while providing the evenly planed surface expected of freshly mown lawns. This is an effect that appears to be easier to achieve as the lawn ages. Perhaps it's due to the architecture and behaviour of the plants having had time to respond to the lawn management regime.

FIGURE 7.25   A frosted lawn. Most species used in T-lawns come from areas that can experience low winter temperatures. Temperatures to −10°C (RHS H4) have so far not been observed to substantially damage T-lawns.

FIGURE 7.26   A closer view of a frozen lawn showing how lawn foliage can catch the frost.

INVASION!

Any lawn, unless it is completely isolated, will be subject to invasion by uninvited species. We know this just by looking at traditional grass-type lawns. After all, most of the species used in T-lawns are those that have demonstrated their ability to survive and thrive in the traditional lawn environment quite uninvited. There are also no surely known limits on how many species a lawn might contain, so we must expect T-lawns to be invaded in just the same manner as grass lawns.

Inevitably, grasses will invade. As outlined earlier, it is relatively easy to remove the larger and more obvious grasses, and this should always be undertaken as part of general and ongoing maintenance, but some of the finer creeping grasses are tricky to spot and tricky to deal with. If you're not up for diligently removing grassy patches and putting in replacement T-lawn plants or new plant tiles, or (provided that you are qualified to do so), applying an effective graminicide; then they are likely to become lawn residents you simply have to put up with (Figure 7.27).

Being new features, there are as yet no examples of T-lawns over a decade old, so how these fine creeping grasses may interact with T-lawns in the long term is essentially speculative. Perhaps, if the conditions are amenable, there is no lawn gardening and with sufficient time, they may eventually overrun the forb-dominated lawn and essentially convert it to a predominantly grass-type lawn, but it also seems quite possible that they will take a place alongside the other resident species and further increase the diversity of plant species in the lawn. Not exactly a pure forb-type T-lawn, but rather a type of grass-managed forb-rich hybrid. For now, we will just have to wait and see.

It is not just a variety of grasses that are likely to pop up uninvited. A variety of other plants have been found spontaneously appearing in T-lawns, some that you perhaps might expect, such as plantains (*Plantago* sp.), and ones you might not, such as *Alchemilla mollis* (lady's mantle). The less-frequent mowing regime, a plant's growth habit and tolerances, and the ongoing removal of highly competitive grasses may be a substantial part of the reasons behind this – allowing some species in, while at the same time keeping others from establishing.

FIGURE 7.27 Invasion to some degree by fine grasses is unavoidable. It is maintenance that determines how successful the invasion may be. Here they have crept in amid the *Mazus reptans*. Keeping a T-lawn perennially and perfectly 'grass-free' is an unlikely scenario without a great deal of care and attention.

Unplanted forb species noted so far, in no particular order, include: *Plantago media* (hoary plantain), *Plantago lanceolata* (ribwort plantain), *Plantago major* (greater plantain), *Plantago maritima* (sea plantain), *Urtica urens* (small nettle), *Taraxacum officinale* (dandelion), *Anagalis arvensis* (scarlet pimpernel), *Veronica filiformis* (creeping speedwell), *Veronica persica* (common field-speedwell), *Veronica arvensis* (wall speedwell), *Medicago lupulina* (black medic), *Trifolium dubium* (lesser trefoil), *Trifolium campestre* (hop trefoil), *Cerastium glomeratum* (mouse-ear chickweed), *Geranium molle* (dove's-foot crane's-bill), *Geranium dissectum* (cut-leaved cranes-bill), *Cardamine pratensis* (cuckoo flower), *Persicaria capitate* (pink bobbles), *Linaria vulgaris* (yellow toadflax), *Dianthus gratianopolitanus* (cheddar pink), *Lepidium didymium* (swine cress), *Erodium moschatum* (musk stork's-bill), *Hypericum humifusum* (trailing St John's wort), *Sagina procumbens* (pearlwort), *Viola arvensis* (field pansy), *Sonchos oleraceus* (annual sow thistle), *Cirsium vulgare* (spear thistle), *Alchemilla mollis* (lady's mantle), *Sherardia arvensis* (field madder), *Scorzoneroides autumnalis* (autumn hawkbit), *Oxalis corniculata* (creeping wood sorrel), *Centaurium erythraea* (common centaury), *Stellaria media* (chickweed), *Euphorbia peplus* (petty spurge), *Cardamine hirsuta* (hairy bittercress), *Rumex obtusifolius* (broad-leaved dock), *Persicaria maculosa* (redshank), *Equisetum arvense* (horsetail), *Cichorium intybus* (chicory) and *Oenothera biennis* (evening primrose).

They mostly tend to come from the immediate locality of the lawns in which they were found, some as prior inhabitants of the lawn's location, or as contaminants in seed from which the plants were grown, with just one or two true mysteries. If this ever-growing list is anything to go by, then there is still plenty of space in T-lawns for more species to fill. If you're not impressed by the invading 'weeds', it may be reassuring to know that many of the above are short-lived in the lawn due to environmental filtering via mowing, and they are also relatively easy to pick out if you have a mind to. However, some just might be tapestry lawn plants in waiting.

CONSEQUENCES

The development of a new type of lawn format is likely to have consequences and probably some unexpected outcomes too. It seems plausible that these will stem from two fundamental aspects of T-lawns: the ideological revision of what a lawn can be and an affiliated botanical outcome.

Once let out of the box and shared, an idea tends to propagate other related ideas. This is especially the case amongst gardeners. Already, T-lawns are in existence that were not created following the researched and tested recommendations laid out in the earlier sections of this book and are garden-type adaptations and variations on a theme; the idea is being propagated and adapted. Although not specifically recommended, these have been included accordingly since they do demonstrate very clearly how once the cat is out of the bag there is no putting it back, and that without a doubt, T-lawns are too young a feature to have the path to creating one set in stone.

However else the concept may influence people, their behaviour and the use of plants in horticultural features is also yet to be determined, but it might be quite exciting. Only a decade or so ago taking a mower to an intentionally planted display of plants in full flower would have been considered horticultural heresy, vandalism even; yet along with the other new plant community approaches to horticulture, it can promote an attractive and for the best part an ecologically benevolent outcome.

Observation indicates how young children with few preconceived ideas on how lawns should be used can initially become completely absorbed in investigating and using T-lawns in ways quite unlike how they treat other garden features, such and flower beds and borders. Like our ancestors, they seem to find scented, flower-filled T-lawns just fine for sitting, laying and playing upon. It is also a fine teaching tool for botany, ecology and horticulture.

BEYOND THE LAWN

The plants and cultivars used in T-lawns may simply not stay contained within them. Just as plants can move into a T-lawn, they can also move out. A move beyond the confines of the lawn would inevitably be facilitated by anthropogenic activities and by a mix of natural processes. The significance of this is likely in part to be related to where in the world you happen to be.

The majority of the plants used in the T-lawn format as outlined in this book are British and European natives that are adapted to the temperate climates of NW Europe and are widespread there. Many are not solely European in their native range and can be found on other continents in the northern hemisphere and exist as imported aliens almost anywhere there are or have been grass lawns. Since they can occur widely and almost wherever lawns can be maintained, it should be no surprise that it will be almost impossible to restrict their movement both in and out of T-lawns.

For the most part this is unlikely to be of note. A common single-flowered white daisy escaping from a T-lawn via seed to another lawn is unlikely to be noticed; it certainly happens all the time between traditional-type lawns. However, T-lawns incorporate distinct phenotypes and selected forms of otherwise common plants. Where selected forms are used, such as the double-flowered *Bellis perennis* 'Pomponette' series, we might expect that until the weaker genetics of the cultivated series are overwhelmed by the stronger and more abundant wild form, that some of the second and perhaps even third generation progeny of wild and cultivar crosses will exhibit characteristics of both types. Through the transfer of pollen and seed, a neighbour's lawn might see the odd pink or double daisy. Since the cultivar is most likely to have originated as a selection from what was originally a wild form plant this should not be overly worrisome, but it does mean we may see some rather interesting related progeny from T-lawn residents turning up in traditional grass lawns despite not having been planted there, and these will not be restricted to daisies alone (Figures 7.28 and 7.29).

Where distinct ornamental cultivars are used that can only be propagated vegetatively, such as *Ajuga reptans* 'Burgundy Glow' (Figure 7.31), the chance of escape seems much less likely, but not impossible. As most gardeners and landscape managers will be aware, plants have a habit of mysteriously showing up in places that are quite unplanned and unexpected.

FIGURE 7.28   An escapee from a T-lawn in a neighbouring grass verge. Although this leaf form was not specifically included in the nearby T-lawn, it is likely to represent part of the genetic heritage of the ornamental white clover plants that were included. From pollen or directly as a seed, the expressed genes have been transferred out of the T-lawn and into the wider white clover community.

FIGURE 7.29   Also, in the same nearby grass verge, another white clover exhibiting the characteristics of ornamental parentage likely to have come from T-lawn plants.

FIGURE 7.30   Not an escapee from a T-lawn, but rather a resident of many years. The regularly wet lawns at Myerscough College, Lancashire, are now home to a large and growing colony of *Lobelia pedunculata*.

A case in question is the T-lawn-worthy south Australian native *Lobelia pedunculata* that has already made it into a few British grass lawns, particularly damp ones (Figure 7.30), and appears to be spreading around the country, rather mysteriously since it is not known to set viable seed in the UK and spreads vegetatively, primarily by rhizomes (although the likely green-fingered culprits may be inferred). Perhaps we may see more of it in the future if T-lawns become more widely known and viewpoints on what constitutes a desirable lawn plant in the UK changes.

NATIVARS

There is currently much discussion amongst ecologists and horticulturists on the role of what have been termed 'nativars', a term coined by Dr. Allan Armitage at the University of Georgia, to describe a cultivar of a native plant that has been selected, crossbred or intentionally hybridised for the purposes of ornamental horticulture [49] (Figure 7.31). Whether a plant is a nativar or not will essentially depend on where you live and to what degree you subscribe to the concept of 'native'. Discussion on nativars is an ongoing topic that reasonably posits that since such cultivars are mostly restricted selections from wild forms that their genetic diversity is not as rich as the wild plants that they have been selected from. As a result, they may be less environmentally robust and of lesser value to the wildlife that is associated with them. Also, since some may be hybrids rather than selections, they may alternatively exhibit hybrid vigour and be stronger than their wild relatives, at least in the short term [50].

This is certainly of some relevance in activities such as ecological restoration, wildlife gardening and plantings undertaken on landscape rather than domestic garden scales. However, investigations into how wildlife interacts with nativars is, so far, inconclusive, suggesting that although pollinators such as bees and butterflies may show something of a preference for wild types, they also have an interest in cultivated types too; it all rather depends on the plant species and cultivar in question. Notably, flies of any kind don't seem to show a preference at all [51–53].

FIGURE 7.31   Recommended as a 'Plant for Pollinators' by the RHS, *Ajuga reptans* 'Burgundy Glow' is a cultivar of a common woodland species native to the northern hemisphere and therefore also a 'nativar' there. It is simply an exotic cultivar of a non-native species in the southern hemisphere.

Since managed lawns are directly associated with humans and they occur in areas of human activity, it seems preposterous to suggest that only wild-type plants are best for use in the domesticated environments in which they are found. However, how the inflow and outflow of genes from ornamental garden cultivars to wild-type plants and vice versa may affect the wider environment continues to remain largely speculative. Although, as the illustrated examples indicate, it is certainly possible that leaf pattern in white clover can be so influenced.

NON-NATIVES

Outside their ecologically suitable ranges and the niches to which they are adapted, most plant populations tend to be relatively short-lived and require continual topping up or a particularly notable influx to maintain any kind of perennial population. Horticulture makes use of many plant species outside of their natural ranges, and far-from-home plants make up a substantial proportion of the plants found in British gardens.

Since T-lawns are a managed garden-type feature, it would be most peculiar indeed to species-restrict them in a way that is not undertaken in the rest of the garden or in British gardens in general. Since it is the plant's suitability for inclusion in T-lawns that is of interest here rather than its ecological point of origin, arrival date or manner of its arrival in the UK, it is just worth noting that well-behaved plants with origins beyond the British Isles can be very useful in tapestry lawns and seem unlikely to bring ruin on the nation's wildlife. Indeed, the study that looked at invertebrates in T-lawns made with both native and non-native species found that mixed-origin lawns were roughly equal to native-only lawns in the number of invertebrates that inhabited them [54]. This broadly matches other research outcomes on the use of mixed-origin plants in garden-type situations [55]. The situation is likely to be somewhat different beyond British shores where the status and behaviour of their non-native plants may be very different. As always, renowned gardener Beth Chatto's advice of 'Right plant, right place' holds true no matter where you are.

8 Tapestry Lawn Plants

Tapestry lawns have been developed in the ancestral home of the lawn, the British Isles and NW Europe. The recommended plants listed here are primarily aimed for use in this region since this is the location and environment where they are known, have been examined and tested, and proved to be useful and generally well behaved when planted in the manner outlined in this book.

The following 31 full plant profiles represent those species that might be considered as 'Tapestry Lawn Stalwarts', in that they have repeatedly been proven to be useful and generally successful in almost all the well-managed tapestry lawns to date. They are by no means guaranteed to be successful in every tapestry lawn, but they represent a good set of plant ingredients to get started with. A list of other species that have also been successfully included in T-lawns but whose performance has been variable and sometimes site specific is also included.

Most of the profiled species are regarded as European natives, although many have extended ranges around the northern hemisphere and are not solely restricted to Europe. Other useful species can also be found in locations with similar moist temperate climates, such as New Zealand, parts of south Australia and NW North America (Figure 8.1). Worldwide, it seems very likely that there will be many untapped species just waiting to creep into use in T-lawns.

Information on how they may behave in T-lawns in other locations around the globe is yet to be amassed, and there are therefore many opportunities for experimentation and assessment of more location and environmentally suitable species in other locales. However, it would seem only prudent to mention again that beyond British and NW European shores that the plants recommended here may be, or may have the potential to be, invasive in other parts of the globe and should be treated with appropriate horticultural caution.

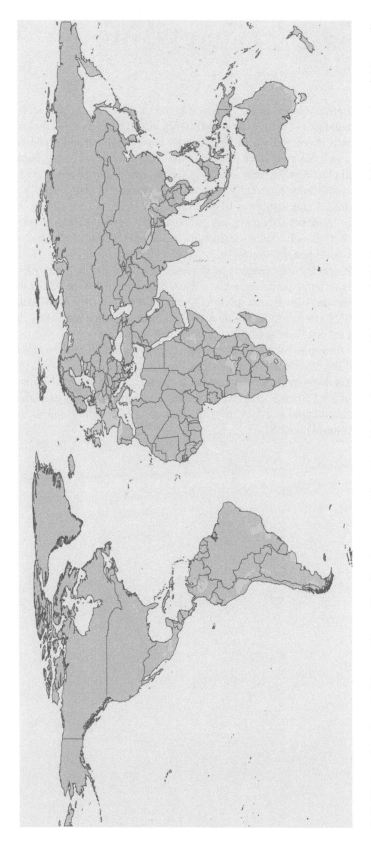

FIGURE 8.1   Based on the Köppen Climate Classification, areas of the globe with broadly similar climatic variables that may be useful in the sourcing of T-lawn suitable plants for use in moist temperate climates. This map is subject to much contention since some of the regions indicated are not climate precise, for example, parts of Chile and the Himalayan region. However, it does broadly indicate where currently useful T-lawn plants have their origins and potentially where others might be found.

Acaena inermis Hook.f.

Purple bidi bidi, purple sheep's burr, purple goose-leaf.

TYPE: All year ground cover. Floral.
REPRODUCTION: Shortly stoloniferous. Seed.
LOCATION: Sun.
SOIL: Well-drained (gritty). Rarely wet for long periods.
USE: *** Many.

Acaena sp. are mostly native to the southern hemisphere, particularly South America. *Acaena inermis* (Figures 8.2 and 8.3) is one of the sub-alpine *Acaena* species from New Zealand. The cultivar 'Purpurea' is the form most frequently encountered and of use in well-drained and sunny lawns. Like most *Acaena*, it can be a long-lived hardy perennial UK (H6) that will spread via rooting stems given the right conditions. In its native habitats, it has both a spiny and a spineless seeded form depending upon the location. The cultivar 'Purpurea' has spineless seeds and colourful foliage. Spiny-seeded *Acaena* are thought to be potentially invasive in the UK and elsewhere, since the small hooked spines are shaped so they are caught in animal hair and can also catch on to clothing and may be unwittingly spread.

It is notoriously difficult to propagate from seed. Many text books suggest the seed should be freshly harvested and then cold-chilled; however, germination remains erratic and poor even if these requirements are met and germination rates are not known to be better than unchilled seed. Found in sub-alpine regions, it is hardy and tolerant of regular rainfall if on well-drained soil, suiting gritty or shallow soils in particular.

IN TAPESTRY LAWNS

The flowers are small and tend not to be in sufficient number within T-lawns to provide a notable display, unless it has formed a distinct patch. It is the coloured foliage that is most useful in T-lawns. It can weave through other plants, doing best when combined with other prostrate and lower-growing species rather than in direct competition with taller species. It does well at the edge of lawns where it can receive most sun. It tends to be a short-lived perennial of 3–5 years, particularly in wet or clay T-lawns, but usually continues to spread slowly via adventious rooting stems. The hue of the leaves is variable ranging from brownish to purple. The leaves are most colourful in sunny locations.

FIGURE 8.2   *Acaena inermis* 'Purpurea' in flower. Leaf colour is variable and thought to be related to levels of soil nutrients and amount of direct sunlight received.

FIGURE 8.3   Coloured foliage is a useful addition in T-lawns, compensating for low floral performance in some T-lawn plants. Here *A. inermis* 'Purpurea' is growing with *Potentilla reptans* 'Pleniflora', a plant it seems to associate well with and *Prunella vulgaris*.

Achillea millefolium L.

Yarrow.

TYPE: All year ground cover. Late floral.
REPRODUCTION: Rhizomes.
LOCATION: Sun. Partial shade.
SOIL: Well-drained.
USE: ** Some.

Achillea millefolium (Figures 8.4 and 8.5) is a native, hardy UK (H6), aromatic rhizomatous peren-nial widely found across the UK. It grows in a variety of conditions and can be found in meadows, roadsides, sparse woodlands, grass lawns, waste grounds, sandy, dry, damp, clay and even saline soils. It is regarded as reliably drought tolerant and used as a lawn substitute in areas that experi-ence frequent drought conditions such as California. It is also tolerant of shading but seldom flowers in shade and may not flower under intensive mowing regimes. It is tolerant of wear and repeated foot traffic. Plants of the species are self-incompatible and require cross-pollination by insects. It is thought to make non-specific mycorrhizal associations that can fluctuate over time, particularly during summer months when it can be in flower. Flowers in the wild form tend to be white or carry pinkish tones.

A. *millefolium* has diverse ecological races throughout the northern hemisphere, occurring in North America, Europe and Asia. These various ecotypes show a variety of forms and environmen-tal tolerances with heights varying between a few centimetres to over a metre. Altitude has been found to influence its growth habit and longevity.

Its rhizomes are relatively slender but increase in diameter with age. Daughter plants are formed at the ends of the rhizomes when they emerge from the soil. Roots are not formed on the rhizomes other than at internodes. New rhizomes are formed at a relatively constant rate throughout the year. In well-watered soil, rhizomes are turgid and brittle. The feathery dissected leaves have a large outline but a small effective surface area, resulting in a decrease in transpiration surface area, and the plant exhibits good drought tolerance. Roots tend to be fibrous although may develop short tap roots. Average root-ing depth is 5–12 cm, with maximum penetration to around 20 cm, and an average spread of roots around 10 cm. Vegetative spread can be rapid under suitable conditions. Seed production is low.

FIGURE 8.4   *Achillea millefolium.* A pale-leaved form, tentatively named 'Golden feather' that is less vigorous than its darker green-leaved counterpart.

FIGURE 8.5   The flowers of *A. millefolium* 'Golden Feather' are reddish-pink.

Sturdy rosettes of plants appear in the first season of growth, with only a small number forming stalks and flowers in the first year. Flowering generally occurs in the second year, although it is possible that a minimum size is required before flowering is possible. Flowering generally occurs from June to September in the northern hemisphere.

Achillea millefolium var. *millefolium* is the form commonly used in ornamental horticulture and is thought to be native to Europe. It may be considered an invasive alien in other parts of the world, including those in which other ecotypes of *A. millefolium* are native.

A. millefolium is long-lived and makes patches that may fragment over time. Mowing reduces the plant's capacity to spread vigorously; vegetative spread is slower than in unmown plants. Leaves may be evergreen over winter and growth is possible in mild winters. It is useful in providing and maintaining winter ground cover if herbaceous species are included in the lawn. Flowering in mid to late summer is not uncommon if mowing frequency is reduced during this period; however, the floral stems are relatively woody and may be unsightly post-mowing. Mowing releases volatile oils in the leaves and produces a distinctive scent that contributes to the T-lawn scent-scape.

A. millefolium var. *millefolium* has long leaves and shows vertical growth when producing floral stems. It is one of the tallest plant species useful in tapestry lawns. It is frequently a mowing trigger species but takes mowing well, largely regenerating between mowing events. It is a useful and robust species that is best planted sparingly, using individuals or small groups of no more than three plants. If tiles are used, half tiles are appropriate. Coloured floral forms can be particularly useful for late summer colour in the lawn. The pale-leaved form is less vigorous and can be planted in greater numbers; it has reddish-pink flowers.

There are yellow-flowered hybrids of *A. millefolium* with Egyptian yarrow (*Achillea aegyptiaca* × *taygetea*), for example, 'Moonshine' that have not been specifically tested in T-lawns, although one or two yellow-flowered forms have popped up in colour-coordinated seed mixes. Also, with the × *taygetea* parentage, is the 'Galaxy' series that includes the RHS Award of Garden Merit (AGM) cultivar *Achillea* 'Lachsschonheit' (Salmon Beauty).

Ajuga reptans L.

Bugle.

TYPE: Floral. All year ground cover. Ornamental foliage.
REPRODUCTION: Stoloniferous. Seed.
LOCATION: Partial shade. Sun.
SOIL: Rarely dry.
USE: *** Many.

Ajuga reptans is an evergreen perennial forb with floral spikes to 20 cm and variable spread to approximately 60 cm. A plant of woodland edges and damp grasslands, it is hardy to zone (UK) 6. It is the commonest of the three native bugles in the British flora, the others being a rare highland species, the erect bugle (*A. pyramidalis*), and the yellow bugle or ground pine (*A. chamaepitys*). Although native and widespread throughout the entire British Isles, it does not set seed in great quantity. The plants propagate mainly by long stolons that root at intervals along their length. With the onset of winter, the runners die, but at every point where leaf-pairs and adventitious roots were formed, there is usually a dormant young plant.

In Britain, the first flowers usually appear in April and can continue until late June or early July. There may be a second flush in some autumns, but this is unreliable, and any floral spikes tend to be much smaller than those produced in spring. The scentless flowers are a purplish-blue, in a spike formed of about six or more layered rings, with generally six flowers per ring. Bugle flowers are hermaphrodite and self-fertile, although the flowers are adapted by their lipped formation for cross-fertilization by bees and contain nectar at the base of the long tube of the flower. The seeds ripen from July to September, but viable seed set is low with embryonic seeds frequently failing to ripen.

Found to grow well in humus-rich, moisture-retentive soils and in partial shade, it can also be found growing in marshy soils. It is suitable for moist sand, loamy and clay soils and tolerates a wide range of pH from mildly acidic, neutral to mildly alkaline soils. For a plant that does well in moist conditions, it also grows in dry deciduous shade and is fairly drought tolerant once established, although when drought stressed the entire plant may die.

The origins of the popular and botanical names of this plant are unknown, although early writers have variously referred to the plant as abija, ajuga, abuga and Bugula. It is thought that the common English name 'bugle', is a corruption of one or other of these forms rather than a comment on the shape of the tubular flowers.

In Tapestry Lawns

There are many bugle cultivars grown in gardens. Several of these are variegated and can be sensitive to prolonged strong sunlight. Exposure to strong sunlight is less problematic for the darker-leafed forms; the form 'Atropurpurea' is thought to be particularly tolerant of full-sun exposure provided that the soil is not lacking in moisture. Most bronze-leaved forms seem to behave in a similar sun-tolerant manner (Figure 8.6).

In T-lawns, most cultivars can be utilised effectively, especially those with coloured and variegated foliage, which can play an important role in the foliage colours added to T-lawns, usually being the source of dark leaves in the lawn. Dark and burgundy leaf forms seem to do better than paler leaved *Ajuga* if the lawn is mostly sunny. Paler-leaved forms are generally less vigorous and may be outcompeted by darker-leaved forms if planted close together.

The exception to this is the small dark-leaved form sometimes listed as *Ajuga* × *tenorii* 'Valfredda' that was introduced into North America (and subsequently the UK) by Valfredda Nursery in Italy under the name 'Chocolate Chip'. It is a hybrid with an Italian small-leaved native *Ajuga* and not a true *A. reptans*. It's performance in T-lawns is variable, sometimes doing well for a period and sometimes not lasting more than a year or so, although other *Ajugas* are seen to do well in the same lawn. If used, this seemingly unpredictable variability should be borne in mind when choosing it. Its floral spikes are pretty but not as noticeable as true *A. reptans*. It is better suited to smaller and less competitive lawns.

All *Ajugas* appreciate some moisture and do best in lawns that don't completely dry out. Growth and spread are much slower in generally drier lawns. An *Ajuga* cultivar that does not suit T-lawns is 'Caitlin's Giant'. Its large clumps look out of place, and it does not respond well to being mown, rarely lasting long.

There are pink and white-flowered cultivars with a variety of habits, all essentially creeping, but with variable vigour and flower spike height. If particularly tall and slightly more impressive floral

FIGURE 8.6   *Ajuga reptans* 'Atropurpurea' and 'Burgundy Glow' in a spring T-lawn.

FIGURE 8.7   *Ajuga genevensis* × *reptans* floral spikes are taller than *A. reptans* and the plant is stoloniferous; however, the plant can be noticeably damaged by mowing and be unsightly immediately post-mowing, although it tends to survive what can appear to be devastating defoliation.

spikes are desired, it is possible to use *Ajuga genevensis*; although this species is a non-stoloniferous plant and mowing can significantly affect its overall look (Figure 8.7).

Useful *A. reptans* cultivars include (Figures 8.8 and 8.9):

'Braunhertz': Brownish red bronzy leaves with floral spikes to 20 cm. Best leaf colour in sun.
'Atropurpurea': A reliable dark-leaved form with blue flowers. Good in sunny lawns.

FIGURE 8.8   'Burgundy Glow' – Variegated tricolour foliage in shades of creamy-white, rose-burgundy and dark green. Blue flowers on medium spikes. Good foliage colour.

FIGURE 8.9   'Pink Elf' – Bronzy-green foliage with soft pink flowers. Tall flower spikes.

'Choc Ice': Pure white flowers with purple-brown leaves and bracts at flowering time.
'Black Scallop': A sport of 'Braunhertz' and the darkest-leaved form, sometimes appearing black when exposed to cool temperatures and direct sun. Blue flowers, generally shorter floral spikes and slow to spread.
'Alba': Pure white flowers with fresh green foliage. Medium-height flower spikes.
'Page's Yellow': Bright blue flowers on short spikes. Bright golden leaves, especially in spring and early summer, but described by one plant reviewer as looking ill.

Argentina anserina (L.) Rydb.

Silverweed.

TYPE: Foliage. Seasonal ground cover. Floral.
REPRODUCTION: Stoloniferous. Seed.
LOCATION: Sun.
SOIL: Moist. Well-drained. Occasionally dry.
USE: ** Some.

Argentina anserina (Syn. *Potentilla anserina*) is a native, hardy, deciduous, rosette-forming, perennial that can be found growing throughout the British Isles across a wide range of soil conditions, from damp and seasonally wet grassy places and beside the sea, to pastureland and on bare ground. Soil pH can be mildly acidic to alkaline. Notably *A. anserina* is only found in sunny locations; it does not tolerate all but the lightest shade. It produces woody, slightly swollen, white roots that develop best in loamy soils, and should you ever feel hungry they are quite edible and reputedly taste of chicken, sweet potato or parsnip.

The form it takes tends to be related to its immediate environment. In places where moisture is rarely limited it tends to grow taller (up to 16 cm) with well-spaced leaf segments and develops with the topside of its leaves predominantly green, although covered in fine silky hairs. In drier environments, the leaf segments tend to be closer, and the fine silver hairs become more apparent on the top side of its leaves and give it its common name of 'silverweed'. It is deciduous, losing its leaves from November onwards when most stolon connections are broken. The plants overwinter as short upright rhizomes with new leaves appearing in April.

It can produce swiftly growing runners up to 1 m long with several stolons along the length. It has been suggested that *A. anserina* uses these as a method to 'escape' from taller or encroaching vegetation. The trailing stems tend to be reddish, and the stolons produced can behave in four different ways. They may be essentially inactive but go on to flower after a year or so, be flowering only, stolon-producing only, or both flowering and stolon producing. Stolon-producing forms tend to be the shortest lived.

It is not a self-pollinating species and must receive pollen from another unrelated individual to successfully fruit and set seed. Pollen from separated stolons (clones) will not result in pollination. Flowers may be male or female and occurrence of either sex may vary on different stolons. Plants produce single, ungrouped, yellow flowers sporadically from May to September. The clear yellow flowers are clearly marked with a bull's-eye pattern, but this is only visible to the ultraviolet sensitive eyes of many insects.

In Tapestry Lawns

Argentina anserina is unusual in T-lawns in that its leaves are clearly silvered by fine hairs, and it is also one of the few deciduous plants that can be used to good autumnal effect, since the leaves turn shades of yellow and brown before withering (Figure 8.10).

During the growing season, it contributes to ground cover, but since its stolons can be wide ranging, it rarely produces large patches – something it can be seen to do in other equitable environments. It tends to be quite mobile in the lawn and once rooted develops according to the amount of sunlight it receives, seemingly doing better at the edges of lawns or where the vegetation produced by the other plants is generally low and light interception is greater. The greater the amount of light it receives the more silver it appears. For a sometimes silver-leaved plant that will only grow in full sun, it is not especially drought tolerant, preferring regular moisture. It tends to disappear from the lawn, retreating to rather unreliable underground rhizomes during extended dry periods.

There is one cultivar: *A. anserina* 'Golden Treasure'. The cultivar originates from two sources in north Wales: The Herb Garden & Historical Plants Nursery on Anglesey, and the garden of David Barrett. It is particularly useful due to its golden-type variegation, especially since although the plant's flowers are attractive to insects, its floral output is low and sporadic (Figure 8.11).

FIGURE 8.10   Autumnal tones from *A. anserina*, here growing with *Viola riviniana* Purpurea group.

FIGURE 8.11 *Argentina anserina* 'Golden Treasure', a red-stemmed cultivar with silvered leaves showing random patterns of golden variegation.

Bellis perennis L.

Daisy.

TYPE: Floral. All year ground cover.
REPRODUCTION: Shortly stoloniferous. Seed.
LOCATION: Sun. Partial shade.
SOIL: Well-drained. Rarely dry.
USE: *** Many.

Bellis perennis is a native short-lived perennial that may behave as a biennial and is found in short turf; it is common in grass lawns but is generally absent from fully shaded and very frequently disturbed habitats. It occurs on a range of soils from mildly acidic to mildly alkaline, with a preference for neutral to alkaline soil. Can tolerate short periods of drought but does best in soils that do not dry out for long periods. Flowering from March to October in the northern hemisphere and may continue to flower through the year if winters are mild. The strongest floral period is between April to June. The flowers are both insect-pollinated and self-compatible. The flower heads of wild forms close at night, and in wet weather, they tend to remain open in ornamental forms. Seed is shed from May onwards. The leaves form a basal rosette and the rosettes remain green through winter and may continue to grow. Short prostrate stolons develop from the base of some leaves to form a small patch. The stolons develop into daughter rosettes with their own root systems. The parent may subsequently die and decay. This type of clonal expansion is relatively slow.

Daisy plants exhibit considerable phenotypic plasticity. Plants from established lawns may be lower growing than those from other grassland habitats due to the process of selection, but there is currently no evidence to suggest that distinct lawn races have developed. There is, however, considerable genetic variation. The foliage is unpalatable due to an acrid secretion, and plants suffer little insect or herbivore damage. An Australasian rust fungus *Puccinia distincta* is found on *B. perennis* as are the rusts *P. obscura* and *P. lagenophorae*; another rust found largely on *Senecio* sp. (Groundsel) may also occur on daisies.

In Tapestry Lawns

B. perennis is one of the most useful of plants for use in a tapestry lawn. They are easy to grow and maintain given the right conditions, and it is almost impossible to have too many.

Since they are short-lived and naturally short in stature and spread, it is rare for them to ever become overly dominant. Although daisies can form small clonal patches, they primarily reproduce via seed rather than clones, so seed must ripen and be shed for the species to maintain a continual presence. Like many short-lived plant species, populations tend to vary over time in response to environmental conditions, and some years may see large numbers of blooms and others less so. The triggers for these variations are not yet well understood. If numbers dwindle it is usually due to poor seed setting and competition from taller growing species such as white clover; oversowing or adding new plant tiles can revitalise their presence. Although largely self-fertile, daisies are visited and pollinated by bees, hoverflies and beetles.

There are a several cultivars commonly available; these tend to be biennial. The colour palette is through white, pink and red and includes double and semi-double forms; although the wild form from which they have been selected is genetically stronger and the colourful double cultivars tend to fade away over time. A lawn laid with all red daisies at its inception can have predominantly white daisies within 3–5 years (Figures 8.12 through 8.16).

Cultivar series include 'Tasso™', 'Pomponette', 'Rominette', 'Monstrosa', 'Bellissima', and 'Habanera'. Individual cultivars include 'Robella'.

There is record of a golden foliage form cultivar known as 'Aucubifolia' being grown and two reputedly purple/lilac flowering varieties known as 'Eliza' and 'Madame Crousse' now lost to cultivation.

FIGURE 8.12 'Tasso' red.

FIGURE 8.13 'Pomponette' mixed.

FIGURE 8.14   'Habanera' white.

FIGURE 8.15   'Robella'.

FIGURE 8.16 Fasciation. The flower bud has been damaged during its development.

Campanula rotundifolia L.

Harebell, bluebell, clychlys deilgrwn, méaracán gorm.

TYPE: Floral.
REPRODUCTION: Rhizomes. Seed.
LOCATION: Sun.
SOIL: Any. Well-drained.
USE: *** Many.

Genetically *Campanula rotundifolia* is an extremely variable species and currently has the taxonomists in a bit of a muddle. However, in the UK where along with pennyroyal it has a different common name in each of the four nations, it is a hardy, rhizomatous, perennial forb that can be prostrate to erect, with long-stalked and roundish basal leaves, narrow stem leaves and blue, bell-shaped flowers. All plant parts exude a white sap when broken that can reputedly be used in a potion for those moments when you wish to turn yourself into a hare.

It has creeping, slender roots that produce adventitious buds. A white tap root is usually present from which rhizomes spread. The tap root tends to be long-lived; however, its leaf rosette rarely survives beyond 2 years and the root is perpetuated by the surrounding rhizomes and secondary rosettes that come from it. Well-established patches can be over three metres across. Plants generally overwinter as hardy green rosettes. In spring, plants make slow vegetative growth that is ground hugging before producing floral stems from June to October. Frequently, the rosette leaves will have withered by flowering time, with new rosette leaves appearing in the autumn. Floral stems are slender and wiry; they generally creep a bit at first, then ascend up to 60 cm.

With a few odd exceptions in the far southwest, the area around the Wash and in southern Ireland, it is a species that is widespread throughout the British Isles, and its range extends throughout most of the northern hemisphere. It tolerates a wide range of conditions, soil types and pH. It can be found growing on coarse sand, calcareous sand, loam, heavy clay, and occasionally pure peat, it is also found in crevices in cliffs and walls. It is generally thought of as a plant of moderately dry and very well-drained soils, although it can tolerate frequently wet locations. It is a sunshine plant that never occurs in deep shade, although it tolerates well-lit partial shade.

Flowering is very variable, with from one to six or more flowers on each floral stem. It is hermaphroditic, but its flowers are largely self-incompatible, and it is insect pollinated, particularly by some bee species such as the Harebell carpenter bee (*Chelostoma campanularum*) that specifically use the pollen of *Campanula* sp. in the rearing of their larvae.

In Tapestry Lawns

With its fine wiry stems and narrow leaves *C. rotundifolia* does not add notably to the ground cover of a T-lawn and is primarily a floral component. However, the growth of its floral stems can be slow at times, particularly during periods of cool weather, and they are frequently mown off before they can be fully appreciated. When they do manage to beat the mower, they are well worth the wait, since the nodding blue bells are held above most of the ground cover vegetation and are clearly visible.

With what amounts to mid/late floral period for a British native it is a useful species for mid to late summer colour. It creeps unobtrusively through other T-lawn plants and is very well-behaved (Figure 8.17). It does best in a sunny location in the lawn. Occasionally the flowers are white.

FIGURE 8.17   The leaf rosettes of *C. rotundifolia* are often hidden; it is its flowers that give it away.

Chamaemelum nobile (L.) All.

'Flore Pleno'.

Chamomile, roman chamomile, noble chamomile, lawn chamomile.

TYPE: Ground cover. Floral. Foliage scent.
REPRODUCTION: Stolons.
LOCATION: Sun.
SOIL: Well-drained.
USE: *** Many.

Chamaemelum nobile (Figures 8.18 and 8.19) is an aromatic, evergreen perennial, hardy to zone (UK) 4. It can grow to 20 cm tall with a spread of around 30 cm. It produces shoots in the first year of growth, which normally form a rosette of leaves, which do not generally flower until the second year. Under the influence of regular mowing it adopts a semi-prostrate, often non-flowering form where the stems creep out parallel to the ground. The single flowers are hermaphroditic, the plant is self-fertile, but it is widely pollinated by bees, flies and beetles. The double form 'Flore pleno' has flowers that are sterile. Although individual stolons may only last a few years, it can be long-lived, propagating itself by clonal spread, and in some populations, by seed.

The double or semi-double flower form 'Flore Pleno' is a result of selection and has been known since the Middle Ages. It is propagated vegetatively by stolons. It is these double and semi-double flowerheads that are specifically grown and harvested commercially for the various chamomile-based teas, and other related products. It is possible to make fresh chamomile tea from plants in your own T-lawn.

It grows across a wide range of pH, tolerating very alkaline soils but doing best on slightly acidic soil. It is often found on light sandy soils but occurs on loams, through to heavy clay soils if well-drained. Although considered a plant of dry, sandy or gleyed soils, a common requirement is to be seasonally wet, usually in winter. On sandy soil, this is achieved by the sun baking the soil hard in the summer, reducing its permeability and leading to temporary waterlogging in winter. It can grow in light semi-shade but does best in full sun and can tolerate short periods of drought.

Despite one of its common names suggesting Italian ancestry, *Chamaemelum nobile* is only native to western Europe (Spain, Portugal, France, the United Kingdom and Ireland), but is present throughout Europe, North Africa and parts of Southwest Asia. Its wild form is listed as 'vulnerable' in the United Kingdom, and it is a species 'of principal importance for the purpose of conserving biodiversity' covered under Section 41 (England) of the NERC Act 2006; it is a UK Biodiversity Action Plan priority species.

IN TAPESTRY LAWNS

If the conditions are right, the combination of good feathery ground cover, a long floral period and one of the most distinctive and pleasant foliage scents for a native species make this an absolute 'must have' plant for a tapestry lawn in the UK. Found growing wild in forb-rich plant communities, it is a seemingly gregarious species and grows well and easily in a sunny T-lawn community. However, the single-flowered wild-type form tends to have erect growth on strong primary stems and will grow tall in T-lawns if they are not mown more frequently than is usually required by the rest of the lawn, making it something of a mowing-trigger species.

If the single-flowered form is used and mown with the same low frequency that the rest of the T-lawn requires, the remaining cut stems can be lightly woody in appearance, rarely recover fully and are generally unattractive. They will remain this way for the rest of the season unless trampled or removed by hand. The semi-double to fully double form 'Flore Pleno' does not exhibit the tendency for strong erect growth and is the preferred choice for a flowering chamomile in a T-lawn.

Although *C. nobile* 'Flore Pleno' may only exhibit a few pale-yellow central ray florets, the flowerheads are much visited by bees, flies and beetles. Since this form is sterile, it is never fertilised and can be continuously in flower from May through to November. The non-flowering form 'Treneague' is also a possible choice for T-lawns as its growth form is continuously prostrate, although this form will not contribute to the floral performance of the lawn.

FIGURE 8.18 The erect, single-flowered form of *C. nobile* can become unsuitably tall in T-lawns due to the infrequent mowing regime. A more frequent mowing regime can induce it to creep rather than grow erect, but this will influence the entire lawn community. Since it is the entire lawn community rather than a single species that is being managed and maintained, it is better to choose a more suitable form to suit the lawn, rather than manage the lawn to suit a single species.

FIGURE 8.19 *C. nobile* 'Flore Pleno' remains low growing under a typical T-lawn mowing regime, and since it is a sterile cultivar, it can be highly floriferous after its first year. Here it is flowering along with *Prunella vulgaris* (selfheal), a plant that it appears to associate well with.

Dianthus deltoides L.

Maiden pink.

TYPE: Ground cover. Floral.
REPRODUCTION: Seed.
LOCATION: Sun.
SOIL: Well-drained.
USE: *** Many.

Dianthus deltoides is commonly called the 'maiden pink'. 'Maiden' since each stem has just one flower to offer and 'pink' is a reference to the fringed petals that usually look as if they were cut with pinking shears. It is a very hardy (UK) H7, evergreen, perennial forb that typically forms loose spreading mats of dark green, narrow, linear to lance-shaped leaves. It is a British native considered to be 'near threatened' since many of its surviving populations are small and threatened by poor grazing practices and scrub encroachment. It is one of the few native species with an RHS AGM.

Single, heavily fringed and faintly fragrant, hermaphrodite flowers appear on branched flowering stems in late spring and summer and thereafter sporadically until September. The wild species flowers close in dull weather and are usually deep pink with a dark, pale spotted band at the base of the petals. They are insect pollinated. Undisturbed it can grow up to 10–15 cm tall and can spread to 60 cm wide.

Dianthus deltoides is a plant of well-drained, mostly sunny conditions usually on alkaline soils overlying chalk or limestone, but also sandy soils and mica-schist, rarely on heavy clay. It can withstand low fertility and droughty conditions, although it does best in fertile, alkaline, gritty loams with good drainage. It can be outcompeted by faster-growing and more competitive species since it spreads relatively slowly. It tends to be found growing in open short grass often with areas of bare soil or rock. Although found growing in dryish conditions, this is usually due to good drainage rather than consistent lack of water. Like most *Dianthus* sp., it does not tolerate prolonged wet soil, particularly in winter. Plants tend to die out in the centre if drainage is not good.

Although predominant reproduction is by seed, *D. deltoides* produces long-lived below-ground stems that initially formed above ground and are known as epigeotropic rhizomes. The oldest stems become covered by soil and litter or pulled into the soil by contraction of roots. This type of vegetative spread is usually slow, only reaching a few centimetres per year. Decay of the initial rhizomes starts usually after a decade or so.

In Tapestry Lawns

A gem in sunny and well-drained lawns or those with shallow or gritty soil. It is a species that has been used successfully in green roofs. In tapestry lawns, it tends to mingle with its neighbours and can be a useful if very discrete green background plant until one of its flowers appears (Figure 8.20).

The floral stems are usually upright and stiffly branched on unmown plants and may be cut off in T-lawns before being seen. The plant tends to respond to having its flowers mown off by producing fewer flowers that appear just above or within the deep green foliage. Within the lawn environment it rarely produces a grand floral show, rather a continuous but slow production of a few flowers over the growing season. Larger patches tend to produce more flowers, particularly when associated with other low-growing and prostrate species that do not act to shade it excessively. To date 'Flashing Light' is the best low-growing cultivar.

There are a growing number of cultivars on the market, usually available as seed, sometimes plants only. 'Albus' has pure white blooms, 'Arctic fire' has small white flowers and a crimson eye (Figure 8.21). 'Brilliant' has vivid red bloom, 'Fanal' has bronze tinted leaves and deep red blooms, 'Flashing Light' ('Leuchtfunk') is the lowest growing form and the most suited cultivar for T-lawns with magenta-pink blooms. 'Red Maiden' has reddish-purple blooms, 'Vampire' has deep-red flowers and dark-green leaves and 'Wisley Variety' has deep-red flowers and dark-green leaves.

FIGURE 8.20   *D. deltoides* 'Flashing Light' growing with *Lobelia pedunculata*.

FIGURE 8.21   *D. deltoides* 'Arctic Fire' with *Thymus praecox*.

Glechoma hederacea L.

Ground ivy, nepeta.

TYPE: Ground cover. Foliage. Floral. Leaf scent.
REPRODUCTION: Stolons. Superficial rhizomes. Seed.
LOCATION: Partial shade. Shade. Sun.
SOIL: Not dry for extended periods.
USE: ** Some.

Glechoma hederacea (Figures 8.22 and 8.23) is a hardy, aromatic, evergreen, low to short patch-forming plant with floral inflorescences up to 20 cm and a spread of around 1 m, with a fast rate of growth. It flowers from March to May and may produce flowers again in September and October, the blue or violet flowers occurring in clusters of three at the leaf axils. The flowers are hermaphrodite, and male-sterile forms of flowers also occur. Despite being self-compatible, its hermaphrodite flowers may still need an insect pollinator to visit for pollination to occur and are frequently pollinated by bees. Many populations of *G. hederacea* produce viable seed, although in some locations, for example, New Zealand and China, seeds are mysteriously not reported. The main root tends to be relatively large with matted smaller roots. It sends out runners up to 90 cm long. The creeping stems can be described as varying from stolons to superficial rhizomes. When crushed or mown the leaves give off a mild mint-like odour. Rabbits do not eat it.

In the UK, *G. hederacea* is a characteristic species of woodland open spaces and edges, hedge banks and scrub communities, although it is also found in open grassland, lawns and in fenlands. It is a highly adaptable species, growing in shade, semi-shade or sun, in a wide range of soils and plant communities. Intolerant of salinity, it is usually found on damp, heavy, fertile, and calcareous soils with a pH range of 5.5–7.5 but also occurs in soils with pH as low as 4.0.

In Tapestry Lawns

A T-lawn stalwart that has proved successful in all tapestry lawns to date. It is noted for creeping through the lawn and providing a useful level of evergreen ground cover throughout the year. The minty odour of the crushed or mown leaves contributes to the characteristic scent-scape associated with T-lawns. Upon mowing, the stolons often fragment and patches tend to be short-lived, although the species does not die out, but rather continually moves throughout the lawn community. It does well in shaded and semi-shaded parts of the lawn, often becoming a perennial resident under and around trees where the soil does not dry out for extended periods.

The variegated form 'Variegata' is particularly useful with its white-splashed leaves. These can act as a partial foil for other flowering plants and provide growing season interest. It is not as vigorous as its green-leaved counterpart and is generally well-behaved, although it shows a tendency to revert to the green form over time. When grown together, the green-leaved wild-type is seen to flower up to 3 weeks earlier than the variegated form, with generally more conspicuous inflorescences and deeper violet-coloured flowers. A combination of the two extends the floral period. When combining the two forms using three tiles of the variegated form to every one of the green-leaved form has proved effective, the green form can go on to dominate in time, frequently due to reversion. The leaves of both forms darken with falling temperatures, and new spring growth may also have darkened tips compared to the rest of the plant.

FIGURE 8.22   *G. hederacea* 'Variegata' is a British nativar that is commonly used as a hanging basket plant and still widely sold under the defunct name of *'Nepeta'*. Its white-splash variegation is a useful foliage colouration in T-lawns.

FIGURE 8.23   Demonstrating the greater vigour of non-variegated forms, the green-leaved form of *G. hederacea* often comes into flower earlier than its variegated counterpart. The floral stems can be stronger and generally more erect than *G. hederacea* 'Variegata'. Floral inflorescences tend to be taller in moist and sunny locations.

Leptinella dioica Hook.f.

Shore leptinella, hairless leptinella, brass-buttons, bachelor's buttons.

TYPE: Ground cover.
REPRODUCTION: Rhizomes. Seed.
LOCATION: Sun. Partial shade. Bright Shade.
SOIL: Rarely dry. Moist.
USE: *** Many.

There is some considerable confusion in the UK and Europe as to which *Leptinella* species is which. Different suppliers and nurseries will supply the same single species under different names. Some species are misnamed entirely, and some are different species names applied to different forms rather than different species. It is confusingly also sold under the synonym *Cotula*. This is a genus that professional horticulture needs to correctly identify and label if it is to be better utilised.

L. dioica is a small fern-like creeping, hardy, semi-evergreen, perennial forb from New Zealand, where it is usually found as a plant of wet coastal habitats. It spreads via rhizomes at or near the soil surface and has relatively small but variably sized slightly fleshy leaves that are usually but not always toothed at the edges, dark green to purple-green and occasionally brownish. A part of the problem in correct horticultural identification is that it is a highly variable species with at least 13 different morphological types, and this has led to its misclassification on several occasions. It may be variable in leaf size and shape, density, vigour, colour, habit, shoot density, leaf length and prostrateness.

It is a member of the daisy family (Asteraceae) and is recorded as both having both male and female flowers on the same plant or being either a male or female plant. Its generally unnoticed furry-looking, button-like flowers are small, 2–5 mm long, yellow-green and slightly longer than they are wide. They are visited by generalist pollinators and will set seed when cross-pollinated; the species is known to intercross with other *Leptinella* sp.

It is very easily propagated from pieces of rhizome and does well in seasonally damp or somewhat poorly drained situations. It is found to perform best where sufficient fertility is available, and the soil does not dry out excessively. Observed to grow on mildly acidic, neutral and mildly alkaline soils, on clay, loam and moist sandy soil, in semi-shade and sunny locations. Coastal locations can suit it well.

In Tapestry Lawns

If this genus had somewhat larger and more colourful flowers, it would be an almost perfect T-lawn plant. As it is, it is a very useful T-lawn plant. Its prostrate and low-growing habit means it acts as a mingling understory plant throughout lawns, with it tending to be overtopped by the taller and larger-growing plant species, something that it seems to mostly tolerate for the periods between mowing events. However, if completely shaded or shaded for a prolonged period by other vegetation, it tends to go brown, lose its leaves and eventually rot away. It can act as a useful textured green background throughout the lawn and fills spaces that might otherwise be empty of plants. It is a good patch filler and may go unnoticed as it slowly spreads throughout the lawn. Its foliage is mildly aromatic, and it contributes to the distinctive T-lawn scent-scape when trodden. Since it rarely grows above 4 cm in height, its scent is not generally noticeable during mowing. If the location of the T-lawn is not especially dry, it is a plant that really should be included.

Its floral contribution is not evident unless you look at the lawn in close-up. It does produce lots of very small dull, greenish-yellow, fluffy flowers that start out looking like stalked and scaly globes and are regularly visited by a surprising variety of pollinators such as beetles, flies and moths. It is not noticeably affected by pests or diseases and is generally the most robust of the *Leptinella* sp. that can be used in T-lawns. When first planting a T-lawn, it is useful to have *Leptinella* plant tiles roughly evenly scattered throughout the lawn, so it can establish a textured green background throughout.

Another *Leptinella* sometimes listed as *L. dioica* 'Minima' is usually *Leptinella minor*; although sometimes it may be either of the other very small-leaved *Leptinella*: *L. filiformis* or *L. nana*. These three very small-leaved type *Leptinella* would make a great addition to a Lilliputian T-lawn. They can be included in normal T-lawns but are usually outcompeted by just about everything, unless in a relatively moist spot where little else will grow due to the shallowness of the soil.

Other useful *Leptinella* sp. are *L. potentillina, L. pusilla, L. squallida* and *L. serrulata*. The cultivar *Leptinella* 'Platt's Black' is also useful, although whether it is a selection from *L. squallida* or *L. serrulata* is currently a matter of debate (Figure 8.24).

FIGURE 8.24 Four *Leptinella*. Top left: *L. minor*; Top right: *L. dioica*; Bottom left: *L. pusilla*; Bottom right: *L.* 'Platt's Black'.

Lobelia angulata G. Forst.

Lawn lobelia, panakenake, alpine pratia.

TYPE: Small floral. Semi-evergreen ground cover.
REPRODUCTION: Rhizomatous. Rooting at nodes. Seed.
LOCATION: Partial shade. Sun.
SOIL: Any. Rarely dry.
USE: *** Many (if you don't have lots of slugs/snails).

Lobelia angulata (Figures 8.25 and 8.26) is a small, hardy UK (H4) spreading, semi-evergreen, prostrate forb with clear white flowers l–2 cm long, on short stalks from the leaf axils. It has the synonyms *Pratia angulata* and *Pratia pusila*. It is native to New Zealand, although generally less hardy forms can be also be found in other locations in SE Asia. It has naturalised in parts of the UK.

Can flower from late spring to early autumn but is generally in flower during the summer months. The plant is not self-fertile. The flowers are dioecious, individual flowers are either male or female, with only one sex found on any one plant. Both male and female plants must be grown in close proximity if seed is required. Pollinated by insects. Flowers are followed by red/purplish berries that may be eaten by birds. Generally considered to be frost hardy but can be subject to the attention of leaf munching slugs and snails, particularly in spring.

Occurs naturally in open or forested areas (sun to partial shade) and is found on both light (sandy) and medium (loamy) soils across a range of soil pH. It can be decumbent in shade or tightly prostrate in full sun.

In Tapestry Lawns

In New Zealand where native snails tend to be carnivorous and slugs eat the algae and microorganisms on leaves but rarely the leaves themselves, it is sometimes listed as New Zealand's most popular native ground cover. It is well-suited to lawns, often found creeping into the garden and parkland lawns of its homeland. It would be a great T-lawn plant if it weren't for the attention it can receive from leaf-munching northern-hemisphere molluscs, most particularly in autumn and spring when it seems to be at its most vulnerable. For currently unknown reasons, when in full growth during summer it receives substantially less mollusc attention.

It is non-dominating and although can spread up to 30 cm in a year it rarely manages to do so. Its clear white flowers continue to appear throughout the summer months after most European natives have passed their main floral period and is a valuable addition to post-solstice colour.

It needs to be planted in large numbers to be significant in its floral contribution and to maintain a presence in most T-lawns. Its berries are generally too small to be noticeable but do add interest to lawns in close-up, since berries in a lawn are quite unusual. The seed is tiny, almost dust like, and difficult to come by.

FIGURE 8.25 *Lobelia angulata* ripe fruits come in shapes known as broad-obovoid to subglobose and look rather like small, partially squeezed berries.

FIGURE 8.26 *Lobelia angulata* re-sprouting in spring after being grazed down to bare stems by leaf-munching molluscs. It is frequently associated with worm casts, but why this might be remains another mystery.

Lobelia angulata 'Treadwellii' (Syn. *Pratia treadwellii*) is generally larger in all its parts. 'Messenger' is more compact, 'Woodside' is a vigorous form and 'Tim Rees' is somewhat larger than 'Treadwellii', with darker green leaves, deep red-purple beneath, and a larger blue-violet flower that was collected in New Guinea and is not considered reliably frost hardy.

Lobelia pedunculata R.Br.

Pratia.

TYPE: Floral. Ground cover.
REPRODUCTION: Rhizomes. Adventitious roots.
LOCATION: Sun. Partial shade. Shade.
SOIL: Moist. Rarely dry.
USE: *** Many.

Lobelia pedunculata (Figures 8.27 through 8.29) is a hardy (UK) H4, evergreen, perennial forb. It is a prostrate plant that creeps across the soil surface or makes small hummocks. It has small-toothed, oblong-to-roundish leaves that can vary between 2 and 13 mm in length and are sometimes a reddish-purple. It readily roots at its many nodes and can create slowly expanding mats of foliage that can be covered almost continuously with small, single, upward-facing, five-petalled, star-shaped flowers from April to October. In addition to rooting at nodes it also produces subsurface rhizomes that extend its vegetative spread. It is found on soil that is rarely dry to constantly moist or usually boggy. It will grow in full sun on consistently wet soils or in partial and sometimes predominantly shaded positions. It is recorded as growing on pH neutral loams but growing particularly well on acidic soils.

It is native to the south Australian mainland and Tasmania. It is notably a highly variable species with the colour of flowers varying from a very pale blue to a deep blue or white. Petal shape and arrangement are also variable. It is a species that could do with further investigation since its variability has led it to be classified under several other names, including *Pratia pedunculata*, *Laurentia fluviatilis*, *Lobelia fluviatilis* and *Isotoma fluviatilis*. It is dioecious, with separate male and female plants, but also recorded as occasionally being bisexual. To date, the plants found naturalised or for sale in the UK are invariably female and seed set is virtually nil. It is frequently visited by butterflies, hoverflies and moths.

Several clones have been collected in Tasmania. The most notable of these is 'County Park' with deep blue flowers, which is also slightly fragrant, and is named after the County Park plant nursery now relocated to Hampshire. 'Tom Stone' is a vigorous and pale blue form but very variable in flower colour. Other listed varieties are 'Buckland Blue', 'Platt's White' and 'Kelsey Blue', although the most commonly encountered forms are 'Blue-star creeper' and 'Alba' (White-star creeper).

In Tapestry Lawns

L. pedunculata is a non-native that is very at home in tapestry lawns. It spreads and mingles very well with other T-lawn plants but does best when not repeatedly shaded out by taller competition. It does particularly well in damp or wet parts of a lawn, whether that is from regular rainfall or a generally moist or occasionally waterlogged soil. It seems to manage compacted soil locations too if it receives regular rainfall. It can wither in full sun if it dries out completely but shows no sun-related problems if consistently moist. In very dry periods it can almost disappear, apparently retreating to its underground rhizomes. It grows fine in partial shade and will even grow in complete (not deep) moist shade. It can be almost continuously in flower between the spring and autumn equinoxes but seems to require a period of warm weather to get going, with a particularly strong floral period in mid and late summer. It is a lovely plant to have in T-lawns, providing continued floral interest after most British natives have passed their prime. It is frequently visited by butterflies.

FIGURE 8.27   In low-growing parts of the lawn *L. pedunculata* (Blue-star creeper) can spread and create a star-spangled carpet.

FIGURE 8.28 *L. pedunculata* 'County Park' has darker blue flowers than its pale blue cousin. It is generally more compact in its growth and slower to spread than 'Blue-star creeper', which may out-compete it.

FIGURE 8.29 *L. pedunculata* 'Alba' (White-star creeper) has generally smaller flowers than either the pale or dark-blue floral forms. It's foliage also tends to be smaller, and of the three forms it tends to be the least vigorous. With it being altogether smaller than either of its cousins, its diminutive flowers appear closer together and it can produce highly floral patches where the flowers overlap in late summer.

Lotus corniculatus L.

Bird's-foot trefoil.

TYPE: Ground cover. Floral.
REPRODUCTION: Seed.
LOCATION: Sun.
SOIL: Well-drained.
USE: *** Many.

Lotus corniculatus (Figures 8.30 and 8.31) is another highly variable species, native to Europe, north Africa and western Asia. It is naturalised in the temperate regions of both North and South America, Australia and New Zealand. It is a hardy, short to long-lived herbaceous perennial and growth form ranges from prostrate to erect. Numerous stems arise from a single basal crown with branches arising from leaf axils. The root system consists of a deep tap root, up to a metre in depth, with numerous lateral secondary roots and a fibrous root-mat near the soil surface consisting of secondary roots and rhizomes, but the plant does not generally spread clonally via rooting rhizomes. Roots make symbiotic associations with specific strains of rhizobium; the resulting root nodules die after mowing and a new generation develops during regrowth.

Flowers usually appear in late April or early May and continue blooming until September. Flower buds and some young flowers can be reddish-brown in colour. Flowers are hermaphroditic but self-incompatible and require pollination via insects. It is visited by butterflies but rarely pollinated by them; pollination is usually by bees. Is a long-day plant requiring daylengths of over l6 hours for full flowering. Seeds are ejected from the long narrow seedpods when they rupture at maturity.

It is typically found in open grassland, well-drained meadows, chalk and limestone downs, hill pastures and montane rock ledges; also, on coastal clifftops, shingle and sand dunes. It is tolerant of all but the most extreme acidic, infertile and poorly drained soils, but intolerant of prolonged shading from other plants, particularly during establishment.

In Tapestry Lawns

Another true stalwart of T-lawns that provides a useful midsummer floral contribution, and so long as it receives direct sunlight, it is rarely fussy in what conditions it is placed. It can be a good choice for difficult spots, particularly if they are dryish and sunny.

Since it does not readily spread via clonal means it usually stays put where it is planted and is rarely problematic as a result. It may require topping up from time to time, but if it is happy with its position it can be long-lived. The roughly circular patches it forms vary in size depending on the conditions in each year, and its floral output tends to be related to the amount of sunshine it receives; with more sunshine overall, more flowers tend to be produced. Some floral forms are a clear yellow in bud and flower, while others start with reddish-brown buds and young flowers that change quickly to yellow as they age. Either works well.

With such a variable species, it is important to select only the prostrate and decumbent forms for T-lawns. The erect form can reach up to 30 cm and is not suited to lawns of any type. There are also commercial varieties specifically used for animal fodder that should not be used, for example, 'Cruz del Sur'. It is entirely herbaceous and can almost disappear during winter, sometimes leaving a sparse-looking spot in the lawn if grown alone or in large single-species patches; its absence is less obvious if other evergreen plants have mingled with it.

There is a double form, *L. corniculatus* 'Plenus' (Figure 8.32), that can be difficult to source and is not available via seed.

FIGURE 8.30   A *L. corniculatus* form with both clear yellow buds and flowers.

FIGURE 8.31   Some forms have reddish-brown young flowers and buds that turn yellow as they age, giving rise to the common name 'Eggs & Bacon'.

FIGURE 8.32   *L. corniculatus* 'Plenus' flowers. A form often described as being dwarf and suitable for rock gardens, it does just as well in T-lawns.

Lysimachia nummularia L.

Creeping jenny, creeping yellow loosestrife, moneywort.

TYPE: Ground cover. Floral.
REPRODUCTION: Stem rooting. Seed.
LOCATION: Semi-shade. Indirect bright light.
SOIL: Moist. Wet.
USE: ** Some.

Lysimachia nummularia (Figures 8.34, 8.35) is regarded certainly as native to Eurasia, including the British Isles, although according to the International Union for the Conservation of Nature (IUCN) Red List it is extinct in Guernsey and Jersey. However, if you live in the north of Britain, the plant you may have growing may not be as native as you think. It is thought that southern plants are more likely to be truly native since although they are not at all floriferous they can and do produce seed, but that northern plants are unlikely to be native since they occur primarily in and around areas of human habitation and although substantially more floriferous than their southern counterparts are not seed producing; they are more likely to have their origins in fragments of the sterile ornamental forms used in gardens that have subsequently reverted from being golden-leaved forms back to being simple green. Even in the south, if the plants are floriferous they probably aren't the wild form. The plant is certainly not native to North America or Australasia but can be found there in the same manner as the forms in northern Britain, growing from fragments but despite many flowers, not setting seed (Figure 8.33).

L. nummularia is a very hardy UK (H6), hairless, low-growing and mat-forming perennial vine that can creep up to 60 cm and roots at nodes. The roots tend to be slender and fibrous. Surprisingly it is a member of the Primulaceae and is very distantly related to primroses. It has solitary five-petalled yellow flowers in the leaf axils (with very small dark red spots if you look hard enough), that are between 1 and 2 cm across and cup shaped. Rarely some plants will produce flowers with six petals and six stamens. It has bright green rounded heart-shaped opposite leaves that can, with some imagination, look like small coins and give it the common name 'moneywort'. It is usually in flower during the day from June to August, though it frequently does not flower at all. It has small seeds that are borne in capsule-like fruits.

It is a plant of open, damp, often clay-rich neutral to mildly acidic soils in shaded woodland and hedges, especially the sides of streams, and in damp grasslands. It is often used ornamentally as a bog garden or waterside plant, as it does well in damp soil and can reputedly survive regular submersion in shallow water.

FIGURE 8.33   A golden-leaved form of *L. nummularia*.

The golden-leaved forms can be eye catching and useful ground cover in T-lawns, particularly in lawns that rarely dry out. Dry lawns tend to see it dwindle within a season or two, and it is best to initially plant it in regularly damp or partially shaded parts of the lawn. It grows surprisingly well with other low-growing plants and rarely forms the monoculture type mats that it can do in uncultivated situations. It does retain noticeable patches for a year or two after planting but then tends to blend in locally. It may revert to green if using the golden-leaved forms in sunny locations.

There are several golden-leaved cultivars variously listed as 'Aurea', 'Gold' and 'Goldilocks'. 'Aurea' has the RHS AGM.

FIGURE 8.34 *L. nummularia* creeping through a T-lawn with *R. repens* 'Buttered Popcorn'. The two plants can often occur together, both doing well in damp soil. The buttercup is held in check for a time by the mower, and the two plants can coexist for a time in a manner not found in unmown situations. The foliage of the golden-leaved form is useful for breaking up the general green background provided by most other lawn plants.

FIGURE 8.35 *L. nummularia* is classed as semi-evergreen but is frequently herbaceous in usual winters. Leaves that remain visible over winter can change colour in response to low temperatures.

Mazus reptans N.E.Br.

Creeping mazus, Chinese marsh-flower.

TYPE: Spring floral. All year ground cover.
REPRODUCTION: Stoloniferous. Rooting at nodes. Seed.
LOCATION: Partial shade. Sun if regular moisture availability.
SOIL: Moist (slightly acidic). Rarely dry.
USE: *** Many.

Mazus reptans (Figures 8.36 and 8.37) is a hardy (UK) H4, ground-hugging, largely prostrate and hardy species that typically grows no higher than 3 cm tall with a spread of around 30 cm. It spreads by creeping stems, which root at the nodes. Produces a dense carpet of foliage that remains green throughout the growing season but may change colour to a brownish purple when temperatures fall below 5°C. May lose its leaves with severe frosts but generally re-sprouts. Small (1 cm), purplish-blue, tubular, two-lipped orchid-like flowers with yellow and white markings appear in small clusters in late spring, generally between May and June, sometimes into July with slighter reblooms later in the year. Observed to grow well with regular moisture and good drainage, it can tolerate excess wetness better than dryness. It is sometimes sold as a bog plant.

It does well in sun if moisture is consistently available, but dry sunny periods can damage it and it may shrivel. Midday shade may be beneficial, and it will bloom well in bright partial and dappled shade. Native to northeast India and the eastern Himalayas through to China, found usually in low alpine valleys and generally moist places. Sometimes wrongly listed under *M. rugosus* (a syn. of *M. miquelii* and *M. japonicus*). Mysteriously, seeds are unavailable commercially, and the plants available are all likely to be clones.

In Tapestry Lawns

Mazus reptans tends to form a lasting patch where it is originally planted if the spot is suitable and spread outwards from it, slowly mingling with other T-lawn plants. Tends not to form strong patches elsewhere. Sensitive to dry periods when establishing and does best in places where it receives a regular supply of moisture either from rainfall or ground-held water. Tends to die out in exceptionally well-drained, dry, hot or overly sunny lawns. Purple flowers are unusual in T-lawns and spreading patches of purple *Mazus* in spring are particularly attractive. Close-up the flowers are thought to resemble little orchids or snapdragons.

The white-flowered form tends to be a weaker grower than its purple-flowered counterpart. The foliage can be somewhat paler, and patch growth tends to be less extensive with less robust spread and generally fewer flowers. Looks best in partially shady spots. Both forms do well in bright but not overly intense sunlight and can benefit from the temporary partial shading effect of relatively taller growing plants in drier and sunnier lawns. Is one of the plants that can have its flowers revealed by mowing.

FIGURE 8.36 *Mazus reptans* purple and white ('Albus') flowered forms.

FIGURE 8.37 Immediately after being mown, the remaining flowers show how the low-growing *Mazus* has spread and weaved its way through the other plants in the lawn.

Mentha pulegium L.

var. *decumbens*.

Pennyroyal, peighinn-rìoghail, brymlys, borógach.

TYPE: Foliage scent. Ground cover. Floral.
REPRODUCTION: Creeping rhizomes. Seed.
LOCATION: Partial shade. Sun.
SOIL: Rarely dry.
USE: ** Some.

Considered to be on the edge of its range in Britain and rarer the farther north you go, *Mentha pulegium* (Figures 8.38 through 8.40) has been cultivated here for centuries, has local names in all four of the main languages of Britain and is considered to have a native form. It is an evergreen perennial mint with creeping underground rhizomes that produce fibrous roots. It is hardy to zone (UK) H7 and flowers in late summer, usually from August to October, with pale lilac-blue to reddish-purple whorled flower clusters, rising as tiers one above the other at each node. The flowers are hermaphrodite and are notably pollinated by bees. It has a characteristic coarse and pungent, mint-like scent to its leaves and is named for its reputed power to repel ants and fleas. *Pulex* is Latin for flea. *M. pulegium* was previously used in homes as a strewing herb, and with the belief that rodents dislike the scent of mint it has also been spread in granaries to keep the rodents off the grain.

It is found to grow well on a range of soils from sandy and loamy through to heavy clay. It tolerates a wide range of pH, showing no specific preference, and will grow in both semi-shade or sun. It does prefer regularly moist and seasonally wet soils.

Although not recognised by major works of taxonomy, British botanists have generally accepted that there are two varieties: an upright form (var. *erecta*) and a prostrate form (var. *decumbens* or var. *pulegium*), with the decumbent variety thought to be native and the erect form introduced. To confuse things further, a particularly robust form has also recently been introduced via seed mixes from the USA. Despite the works of taxonomy, DNA analysis confirms that there are indeed two forms, which may breed and form intermediates where they grow together. Although var. *erecta* is often associated with known introduced sites, it is not always the case, and both forms could equally be native in some localities.

The erect form is considered a problematic weed of cultivation in much of the rest of the world; however, in the UK the (native) species is currently considered 'endangered' after being downgraded from 'vulnerable' in 1999, its rarity status here being explained by the distinguishing of native and non-native populations. It is a 'Priority Species' in the UK Biodiversity Action Plan, and it is protected on several Sites of Special Scientific Interest (SSSIs).

In Tapestry Lawns

One of the most distinctively scented plants for use in T-lawns, it readily releases its coarse minty scent when walked upon and when mown. Not ideal for notably dry lawns, but if the lawn or the part of it you plan for the pennyroyal does not frequently become bone dry, it is a species you really should try to include. Along with chamomile, the leaf scent is one of the most notable to be found in T-lawns and significantly contributes to the T-lawn scent-scape.

It tends to do best if not in direct competition with larger-leaved plants but seems quite able to move through the lawn and will often make distinct, if temporary, patches. Although the flowers are not a major part of a lawns' floral performance, they are very attractive to bees and other pollinators. Cut stems tend to turn black and may be removed, but they do not generally affect the overall aesthetic.

An unusual observation, perhaps worthy of more investigation, is that very large, pale slugs with an orange fringe, potentially *Arion* sp. (which includes the introduced Spanish slug *Arion vulgaris*), seem to be attracted to prostrate pennyroyal patches, and if they are present in the lawn, can often be found stretched out upon them in wet weather. Oddly, they do not appear to be consuming the plant. It makes them quite easy to pick off and dispose of.

FIGURE 8.38 Prostrate pennyroyal is a ground-hugging mint that forms distinct patches and also mingles well with other low growers, here with *Acaena buchananii*. It does best in regularly moist lawns. Cut pennyroyal stems have a tendency to turn black, but this low-growing variety generally creeps below the 4 cm cut height applied to lawns, and this is rarely a problem.

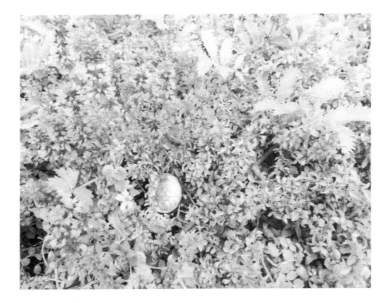

FIGURE 8.39 *M. pulegium* var. *decumbens*. A useful ground covering and widely spreading form that has floral stems to 8 cm much visited by bees. When trampled or cut, the distinctive and strongly mint-scented leaves add another sensory dimension to the lawn.

FIGURE 8.40 *M. pulegium* var. *erecta*. A much taller growing form that is unsuitable for T-lawns. It rarely develops good ground cover due to its erect growth habit, and the stems turn black and woody after mowing.

Parochetus communis D. Don

Blue oxalis, blue pea, shamrock pea.

TYPE: Growing season ground cover. Late floral.
REPRODUCTION: Stoloniferous rooting at nodes. Seed.
LOCATION: Partial shade. Sun.
SOIL: Rarely dry for long periods.
USE: ** Some.

Parochetus communis (Figures 8.41 and 8.42) is a prostrate, deciduous, perennial forb with slender stems, rooting at the nodes. It is hardy, and unmown, its leaves can grow between 10 and 20 cm tall. Its leaves are trifoliate (with three parts and akin to that of a cloverleaf), and they can be plain green or marked with brown-yellow variegation.

The sweetly scented flowers of *P. communis* are borne usually singly or occasionally in clusters of up to three flowers on stalks in late autumn. The flowers are generally a clear true blue; occasionally, white or purple and pink has also been recorded. After pollination, seeds occur inside brown pods sometimes forced open by seeds germinating inside. It is native to the low Himalaya regions and other Asian mountainous regions; surprisingly, it is also native to some Afro-tropical mountains. It generally grows in damp, shady places on forest floors or on the banks of streams and rivers at altitudes of between 1500 and 3000 m.

A rather complicated plant first recorded in Nepal in 1825 by David Don and then collected in Africa for the first time in 1871 on David Livingstone's Zambezi Expedition. Two populations, one in mountainous tropical Africa and one in the low Himalayas and SE Asia, have led to some discussion amongst botanists and gardeners as to whether they are truly the same considering their origins and consequently which might be best to grow in our temperate climate. The Himalayan type is seemingly much preferred amongst alpine enthusiasts since it is considered the hardier of the two types, although tropical African mountains experience both frosts and snow. The African form is thought to be the most frequently available in the UK and can occasionally be found for sale in the autumn due to the species' habit of flowering between October and November.

It is sometimes recorded that in the wild, *Parochetus communis* can develop rhizomes and sometimes produces tubers up to 4 cm long, although these remain rarely recorded. At one time, the forms had two species names *P. communis* and *P. africanus*. It is frequently suggested (across the garden fence and in occult horticultural lore) that the two populations can be identified by the markings on the leaves with the Himalayan form having darker and clearer markings, also that the overall habit can give the origins away, with the African form having a strongly creeping habit and the Himalayan less so. There is, however, notable natural variation in all the identifying features of the species whatever their origins, and thus provenance has become particularly relevant to some horticulturalists. However, the consensus now is that the African form is the subspecies *P. communis* subsp. *africanus* and the variations in overall hardiness remain poorly validated since forms of either origin can expire unexpectedly in what should be ideal conditions.

The species is in the Fabaceae botanical family (a.k.a. the Legume family, along with peas and beans) and the Trifolieae tribe, along with other three-leaflet plants like clovers, although it is considered significantly different enough from clovers that where it sits has been jiggled into its very own subtribe Parochetinae and forms a monotypic (one of a kind) genus. Like clovers and other legumes, it develops clear rhizobial associations with bacteria in its roots and can often be found with tiny grey root nodules that may well be an occasional source of nitrogen for the lawn when it is mown.

In Tapestry Lawns

True blue is generally a rare flower colour in the plant world and to be able to include it in T-lawns is an opportunity not to be missed, particularly since the flowers generally appear after the autumn equinox from mid to late autumn when floral performance in the lawn is winding down and largely left to daisies and autumn-flowering geophytes such as the autumn crocus. The frequently variegated leaves are a valuable ground cover through the growing season and will confuse many who will misidentify it as a clover. It is deciduous and the leaves are frost sensitive, tending to collapse shortly after the first frost. Its naked stems will remain after the leaves have fallen away, and it is best planted to mingle with creeping evergreen species for best winter cover.

In good years, *P. communis* can produce patches of blue flowers that draw the eye and attention in the last floral bloom of the year, although the performance varies from year to year and the weather can be critical. The conundrum here is that although both forms are contentiously thought to be hardy, *P. communis* flowers and leaves will succumb to ground frost, and therefore floral performance is best in years when the first frost arrives late. After waiting all year, it is possible to see the blue floral patches appear and then suddenly turn to mush overnight, particularly in exposed locations; sheltered spots and urban areas tend to fare better. Perhaps with global warming, the chances of having a blue autumn bloom will improve.

Along with *Clinopodium douglasii*, *P. communis* can sometimes respond poorly to being moved from where it was propagated to where it is eventually laid in the lawn. This transplantation shock can result in the plants leaves being lost or in little or no growth in its new position. This is usually temporary, although it can take a month or more for the plant to sufficiently recover to regrow. If this should happen in your newly laid lawn, take it as an opportunity to sprinkle a few lawn-tolerant therophyte seeds such as *Viola tricolor* or *Anagallis arvensis*; they should help fill the gap and perhaps bring a bit of extra welcome colour while anticipating that first frost.

FIGURE 8.41   *P. communis* showing the ringed marking on its trifoliate leaves. Garden lore has it that the clearer the markings the more likely that the form is Himalayan. Since on both the African and the Himalayan form leaf markings are highly variable, this is not a reliable determinant.

FIGURE 8.42   Patches of *P. communis* in bloom in a new T-lawn. November, Southern England.

Pilosella aurantiaca (L.) F.W. Schultz & Sch. Bip.

Fox & cubs, orange hawkweed.

TYPE: Ground cover. Floral.
REPRODUCTION: Stolons. Seed.
LOCATION: Sun.
SOIL: Well-drained.
USE: ** Some.

With the synonym *Hieracium aurantiacum*, Pilosella aurantiacum (Figures 8.43 and 8.44) is one of those plants that just didn't make it across what was to become the English Channel before the sea spilled in, probably since it was to be found growing up to 2500 m in the mountains of central and southern Europe. However, despite a general lack of alpine conditions in the UK, within a hundred years of its arrival as an ornamental plant in British gardens (circa 1629), it had escaped, naturalised and has for all intents and purposes become part of the modern British flora. Arguably it might even be what is sometimes known as a 'near native', not only because it comes from quite nearby geographically speaking and has plant cousins here, but also since it can now be found in every county across the land; although rather mysteriously for a true alpine, it tends to remain uncommon in British mountains, finding reason to prefer our lower altitudes. Research undertaken in Japan suggests that the presence of other vegetation is more important than soil characteristics in determining orange hawkweed establishment, so perhaps the particular mountainous conditions and vegetation of Britain are in some way unwelcoming to this fellow alpine.

In the UK at the moment, there are two disputed subspecies, *P. aurantiaca* subsp. *aurantiaca* and *P. aurantiaca* subsp. *carpathicola*, with the latter being thought to be the most abundant. However, it is also postulated that subsp. *carpathicola* doesn't even really exist at all and that they are all subsp. *aurantiaca* expressing, amongst other things, floral colour variation from a rare yellow through to pale and deep brownish-orange. Yet another of those botanical riddles. And to add to it all, older reference materials will usually have this plant listed under its synonymous scientific name of *Hieracium aurantiacum*.

If your brains have not quite turned to mush and dribbled out of your ears, you will be pleased to know that the plant described here is simply *Pilosella aurantiaca*, commonly known as fox & cubs. The common name is a reference to the way that many of the unopened flower buds cluster underneath those that have opened, rather like fox cubs, kits or pups nuzzling their mother (botany isn't the only subject that can't quite make up its mind on names). The usual fox-orange colour and abundant black hairs complete the picture. Its other common name 'orange hawkweed' is a simple description of the flower colour and that country lore suggests hawks will swoop down to rip off and eat the elevated flowers to improve their eyesight, in the same way carrots are supposed to do for humans.

P. aurantiaca is a hardy H6 (UK), monocarpic, stoloniferous and shallowly rhizomatous perennial forb that exudes a milky sap when damaged. It is found growing in gardens, railway and roadside banks, on walls and in churchyards and other grassy and unutilised open places, usually in a mostly sunny position. At the base of each plant is a shallow, fibrous rooting rosette of hairy oval-to-lanceolate leaves, from the centre of which rises the floral stem, bearing a tight cluster of up to 12 or more flowers that are between 1 and 2.5 cm in diameter. The floral stem is often described as short but can reach over 40 cm, seemingly responsive to the local conditions. In the UK it generally produces its first flowers in June and continues through summer and into late autumn. The stolons continue to elongate during the summer months, forming new rosettes at their tips. The stoloniferous connections break as roots anchor the new rosettes, and the new young plants become independent. After flowering, the supporting rosette dies.

It is a plant with few soil associations, being found on almost all soil types, but particularly loams; however, there appears to be a commonality in that most conditions it inhabits tend to be well-drained, well-textured and moderately low in organic matter. There is some thought that it may have allelopathic properties, but these are yet to be usefully substantiated beyond the laboratory.

In Tapestry Lawns

The mid- to late-summer flowering habit of *P. aurantiacum* and its virtually unique colour is a welcome feature in T-lawns. It is considered to be a bit invasive in some garden borders, since it may spring up from seed and spread by stolons and shallow rhizomes, but like its cousin *P. officinarum*, in T-lawns it tends to be much less prone to take over or spread unmanageably; rather, it weaves its way through the other vegetation and may even find the competition too strong to withstand in some places with patches being short-lived. This is likely to be partially due to the monocarpic nature of the flowering rosettes but also a response to the height and density of its neighbours. It is seen to do better in generally low lawn vegetation that rarely meets the 4 cm cut height, rather than with plants that can substantially overtop its own leaves for any length of time. It behaves as if it likes sun on its hairy leaves and will lift them as high as it can if other plants encroach upon them. It may also creep away from shaded areas using its stolons to seek out brighter spaces.

Although like many patch-making hawkweeds it has a reputation for being allelopathic and poisoning nearby plants; this is a poorly substantiated reputation and allelopathy has not yet obviously shown itself in T-lawns. Depending on its neighbours it is usually well-behaved if not a lot weaker than might initially be imagined.

Observing it in flower can sometimes be tricky since each rosette usually puts up only one or two floral stems, and if they happen to be mown off before blooming, then those flowers are lost. It may produce additional floral stems that can appear as the summer growing stolons continue to elongate, although it tends to be the much later flowers that are usually seen in any numbers. It is fortunate it can continue to flower into October and sometimes November. It is in these later months of the year that the non-natives incorporated in T-lawns can come into their own and continue to provide both colour and floral resources to foraging insects.

FIGURE 8.43   With the onset of midsummer, the mowing requirement decreases, and it is possible to allow some taller species like *P. aurantiacum* to rise above the rest of the lawn.

FIGURE 8.44   Standing out in midsummer with *Achillea millefolium*. These relatively tall flowering species do not have to induce mowing and offer useful summer colour in the lawn.

Pilosella officinarum F.W. Schultz & Sch. Bip.

Mouse-ear hawkweed.

TYPE: Ground cover. Floral.
REPRODUCTION: Stolons. Seed.
LOCATION: Sun.
SOIL: Well-drained.
USE: ** Some.

Pilosella officinarum (Figure 8.45) is another of those plants that seems to have multiple botanical identities, having received eleven other Latin names, and having twice been named *Hieracium pilosella*. It is indeed a 'hawkweed' (*Hieracium* from the Latin for 'hawk'), but the genus *Hieracium* is subdivided into three sections, one of which is *Pilosella*. In these sub-divisions, a notable feature of true *Hieracium* is that they do not produce stolons, and like *Pilosella aurantiacum*, *Pilosella officinarum* most certainly does. So, Mouse-ear hawkweed is indeed a *Hieracium*, but more accurately it is a *Pilosella*. Just to make it even more 'interesting', it is also a very variable species with quite a few subspecies and forms (although mercifully only seven are thought to be found in Britain), and hybrids with other related species are also known to occur.

It looks similar to a dandelion when in flower; however, the flowers are a lighter lemony yellow and the shorter smooth-edged rather than toothed leaves are covered in fine hairs. It is a sun-loving, perennial forb that can grow to 20 cm tall with an equal or greater spread and is fully hardy to zone (UK) H5. It usually develops flat rosettes linked by a slender rootstock. New rosettes are produced seasonally on stolons, in the leaf axils of senescing rosettes or from the underground bud bank on the rootstock. Stolons grow up to 30 cm long, are sometimes branched and usually have a rooting, terminal rosette of overwintering leaves. The number, length and degree of branching of the stolons depends on the population density. It can form patches that exclude other plants, and it has been suggested that it may have an allelopathic effect and reduce the growth of other plant species nearby since it contains a number of phenolic compounds with phytotoxic properties, and leachate from the roots has been found to inhibit the germination of its own seeds.

Rooted stolons form on rosettes that are due to flower. Rosettes may take from one to four or more years to flower. It is monocarpic and within a few months of flowering the parent rosette senesces and the stolons decay, leaving independent rosettes. The new rosettes may be initially sensitive to frost but mortality is generally rare.

Basal rosettes have leaves with a whitish tinge underneath but are generally greyish-green with long white hairs. The plant gets its common name from these hairy mouse-ear-shaped leaves, although sometimes you have to imagine something of a mutant mouse-rabbit hybrid. A single erect floral stem arises from the rosette with a single lemony-yellow flowerhead, the outer florets often red striped beneath. Sheep and rabbits seem to rather enjoy the taste of them, but not the leaves. It is generally in flower from May to September, with peak flowering occurring in June with a second flowering in late summer. The flowers are hermaphrodite but are self-incompatible and are pollinated by bees and other insects. Seed is produced both apomictically and by insect pollination. Apomictic seeds produce plants that are essentially clones of the parent. Seedlings rarely develop into adults in established vegetation and may remain in a juvenile, non-rosette form for several years. A number of butterflies, moths, gall flies and gall wasps are associated with this plant and will specifically feed from the flowers or lay their eggs on the leaves or stems.

It is found growing on a wide range of soils from sandy and loamy through to heavy clay, and through a range of pH and moisture conditions from acidic to alkaline and moist to dry. It is found to grow particularly well on well-drained and nutrient-poor soils and is most common on dry sandy soils where its reputation for being able to recover from apparent desiccation is well-warranted. Shetland mouse-ear hawkweed (*Pilosella flagellaris*) is similar but has two or more flowers per floral stalk, and as its name would suggest it is found in the Shetland Islands.

In Tapestry Lawns

Despite its allelopathic reputation in grasslands, in tapestry lawns *P. officinarum* appears to behave quite socially, behaving more as a mingling creeper than as a mat-forming plant. Its flowers tend not to appear until June onwards, some weeks after the 'Chelsea chop'* in May. In grassland stands, the flowers can be quite noticeable since they are usually in dense groups; however, the creeping habit it appears to adopt in T-lawns tends to mean the flowers are usually well-spaced, only occasionally appearing in substantial clusters.

FIGURE 8.45   *P. officinarum* in an early summer T-lawn with *Bellis perennis* 'Robella' for company.

The clear lemony-yellow, almost sunburst-type flowers of *P. officinarum* can add height to a T-lawn without making a canopy and are no reason to get out the mower. Along with late-flowering *P. aurantiaca* and *Achillea millefolium* the effect can be something akin to tall coloured parasols in the mid- to late-summer lawn. Usually each rosette cluster will only produce a limited number of flowers before senescence sets in and the height of these can be variable. The variety 'niveum' has notably silvery leaves and is useful in dry sunny T-lawns.

* Chelsea chop – the cutting back (in T-lawns usually the first mowing) of herbaceous plants around the time of the RHS Chelsea flower show.

Potentilla reptans L.

Creeping Cinquefoil.

TYPE: Foliage. Ground cover. Floral.
REPRODUCTION: Stolons. Seed.
LOCATION: Sun.
SOIL: Well-drained.
USE: ** Some.

Potentilla reptans is a hardy (UK) 5, native perennial growing to 1 m (3 ft 3 in) by 1 m (3 ft 3 in) that creeps widely across the ground and mingles with other plants. It has deep roots and five-fingered leaves that occasionally have seven fingers. It has a Eurosiberian range and is found commonly across the UK but is rarer in Scotland where populations are thought to have been introduced. It has far-creeping stolons up to 1.5 m long that are in flower from June to September. The bright yellow flowers have five rounded petals that partially close in dull weather and fully at night. The flowers are hermaphrodite and self-fertile but are pollinated by bees, flies and beetles. It does not grow well in shade.

Common to woodland rides, open grasslands, hedgerows, banks and roadsides, and waste and cultivated ground, generally on neutral to alkaline soils, although it is found on all soil types that are well-drained. It might be considered to be something of a clever plant in that it has been used in experiments published in the journal *Nature Communications* that have determined that it can evaluate the density and competitive ability of its neighbours and tailor its responses accordingly.

In Tapestry Lawns

A plant that is intermediate in its ability to offer ground cover and flowers, *P. reptans* is nevertheless a useful and welcome plant in T-lawns; it fits all the criteria for a T-lawn plant and should be included in all T-lawns if possible. Considering the irregularity and low number of single flowers produced, it is surprisingly attractive to pollinators and is useful in its ability to mix and mingle with other plants.

It can form open, wide intermingled patches amongst the other plants, but these tend to be highly mobile and short-lived. It is particularly useful in spots where little else thrives. It seems to root well in otherwise difficult situations such as compacted soil, and if the long stolons remain connected, with one stolon in a moist spot, it manages to survive dry locations. There is a useful double-flowered form 'Pleniflora' (Figure 8.46).

FIGURE 8.46   In sun it has reddish stems that would be a notable feature in other ornamental plants, and the double form 'Pleniflora', also known as 'Royal Cinquefoil', has larger and more noticeable flowers that will speckle a lawn with flowers over the summer season. Generally, trouble free.

Primula vulgaris Huds.

Primrose.

TYPE: Ground cover. Floral.
REPRODUCTION: Seed.
LOCATION: Sun.
SOIL: Well-drained.
USE: *** Many.

Primula vulgaris is one of five native primroses and is widespread across the British Isles. Its British and European distribution is likely to be related to moisture availability and atmospheric humidity, and it is less common in parts of Europe with continental rather than maritime climates. Prolonged dry conditions are thought to be detrimental to the species, and parts of East Anglia and southeast England have seen numbers declining.

It is a shade-tolerant and shallow-rooted plant of moist open grasslands, woodlands and hedgerows. It is found across a range of soil types from mildly alkaline to mildly acidic and grows particularly well in heavy clay soils and damp shady habitats where moisture is rarely unavailable. It is a long-lived, hardy (UK) H6 evergreen perennial that produces hemispherical clumps of rosette leaves from a stout rhizome over time and may spread to 30 cm. Unlike most T-lawn plants it does not notably spread clonally but instead produces new rosettes that remain attached to the mother plant and is propagated by seed.

In mild winters it may be in flower from December but is usually at its best from March to early May with seeds ripening between April to August. Seeds have an oily elaiosome that attracts ants. The delicately scented pale-yellow flowers have a deeper yellow or orange-yellow centre and are held above the plant on short slender stems, though in large populations white and pale pink forms occur. The flowers are hermaphrodite (with both male and female parts) and of a type known as heterostylous, where individual plants have either 'pin' (style prominent) or 'thrum' flowers (stamens prominent). Fertilisation takes place between pin and thrum flowers. A rare third type with both pin and thrum in the same flower is occasionally found in the SW of England.

The flowers are visited by early flying bees, particularly bumblebees, in addition to a wide variety of flies and both brimstone and small tortoiseshell butterflies. There are two notably unusual ornamental forms: 'Hose-in-hose' where one flower appears to grow out of another and 'Jack-in-the-green' where the flower is surrounded by a green leaf-like ruff.

In Tapestry Lawns

A great harbinger of spring in T-lawns producing dense clumps with leaves that will usually grow above the cut height of 4 cm (Figure 8.47).

Almost inevitably when *P. vulgaris* plants are mown they will be almost completely defoliated (Figure 8.48). However, the plant will usually have stored sufficient carbohydrate in its roots to recover, and the new rosette of leaves produced subsequently usually remains lower growing and less prone to the mower until the following year.

Most florally effective when lots are planted in scattered groups across the lawn. Groups of five or more produce the best results. The many-coloured hybrids available from garden centres can also be used, but these tend to be much shorter lived. Whereas a true *P. vulgaris* may last for over 20 years, the garden hybrids tend to be bred for a one-time spectacular performance and may disappear after a few years. The garden hybrids are also, depending on variety, seemingly more attractive to slugs, rabbits and birds that will nibble the petals down to the calyx and ruin the overall display. Purple and mauve varieties seem particularly prone to being munched. The generally evergreen leaves are useful cover in sparse winter lawns, maintaining touches of green amid the darkening hues of many of the other species.

Although more usually associated with woodlands and shady conditions, *P. vulgaris* is found in the wild in open and exposed places. With sufficient moisture provided by rain or retained in the soil, it seems quite at home in the open environment of a lawn and has been seen to do well year after year in completely sunny lawns if they only rarely become dry.

FIGURE 8.47   *P. vulgaris* leaves can grow tall in tapestry lawns.

FIGURE 8.48   A freshly cut *P. vulgaris* showing how mowing can almost completely defoliate the plant.

Prunella vulgaris L.

Selfheal.

TYPE: Floral. Ground cover.
REPRODUCTION: Seed. Adventitious roots (short surface rhizomes and stolons).
LOCATION: Partial shade. Sun.
SOIL: Well-drained.
USE: *** Many.

Prunella vulgaris (Figures 8.49 through 8.51) is a hardy (UK) H6 native and usually perennial forb, with often square, crimson-tinged stems that can be erect to decumbent, spreading to around 20 cm and reaching up to 30 cm tall. Depending on location it is usually in flower from May to September, and the seeds ripen from August to October. The flowers are purplish-blue, rarely pink, or white; are hermaphrodite, self-sterile and are pollinated by insects, particularly bees.

P. vulgaris occurs widely across the British Isles and is found mainly in meadows, pastures and field margins in short or open turf, except on the most acid of soils. It can tolerate shade but is usually found in lightly shaded to mostly sunny locations on level ground, showing a preference for soils that rarely dry out. It tolerates a wide range of soil types from light sandy soil through to heavy clay and a wide range of pH values. Due to its bitter taste, herbivorous mammals such as rabbits and deer do not generally eat it.

Overwintering as a rosette of leaves, new shoots elongate in late spring. The creeping vegetative branches produce adventitious roots when in close contact with moist soil. Clonal expansion can be relatively fast although the shoots tend to die after flowering, which can cause the whole plant to die, unless there is basal regeneration. In favourable environments, many rooted shoots with short internodes are produced, creating patches. Some plants produce only flowering stems; a few may have only vegetative shoots and others a mixture of the two. The proportion of shoots that remain vegetative or produce seeds is genetically determined. *P. vulgaris* exhibits clear phenotypic plasticity, and its growth habit is notably affected by mowing intensity and environmental conditions. In lawns, it usually has a prostrate habit.

Botanists often refer to two varieties of the species, the Eurasian one being called *P. vulgaris* subsp. *vulgaris* and the North American one, thought to be native there, being referred to as *P. vulgaris* subsp. *lanceolata*. By flower they are almost indistinguishable, although in form subsp. *lanceolata* is usually more upright, sometimes reclining on the ground with tips ascending and with narrower mid-stem leaves than the usually prostrate to decumbent subsp. *vulgaris* that has wider mid-stem leaves. This distinction may be relevant for T-lawns on different continents if 'nativeness' is a consideration. In continental Europe, both annual and perennial races are known to occur.

In Tapestry Lawns

P. vulgaris subsp. *vulgaris* is a true stalwart of British T-lawns, being a successful and enduring species in all T-lawns to date, and is more appropriate in T-lawns than subsp. *lanceolata*. It provides both deep green ground cover throughout the year that is particularly useful during winter, and it reliably produces a good floral display.

Selfheal plants can respond to mowing by producing new growth that is more prostrate; however, there are tall phenotypes of subsp. *vulgaris* that are unsuitable for use in T-lawns. Unsuitable forms tend to produce taller, leggier stems and long inflorescences with small flowers, and are often found in 'wildflower seed'. Some suppliers offer pink and white forms, usually as plug plants, and although pink and white forms of *P. vulgaris* do occur, the forms offered as plug plants are usually mislabelled forms of *P. grandiflora*.

FIGURE 8.49   A prostrate form of *P. vulgaris* subsp. *vulgaris* growing to about 3 cm tall with relatively large flowers on short inflorescences. Forming patches, the purplish-blue of the flowers is most noticeable during summer.

FIGURE 8.50   *P. vulgaris* subsp. *lanceolata* generally grows too tall for use in T-lawns.

FIGURE 8.51   A white-flowered form of *P. vulgaris* subsp. *vulgaris*.

Ranunculus repens L.

Creeping buttercup.

TYPE: Ground cover. Floral. Ornamental foliage.
REPRODUCTION: Stolons. Seed.
LOCATION: Partial shade. Sun.
SOIL: Moist.
USE: * Few. Individual scattered plants, ideally only cultivars.

The British amongst many others have something of a love-hate relationship with this easily rec-ognisable native. Generations of gardeners have spent much time and energy keeping it out of their gardens, tolerating it sometimes only in lawns, and preferably those that belong to someone else. It is however quite a distinctive plant and generations of children have held the flowers of buttercups under their chins to confirm their liking for butter, and buttercup-filled meadows have a place in the British psyche, although the buttercups there are more likely to be the taller-growing meadow buttercup (*Ranunculus acris*). If it were not for its ability to spread so easily, and in almost any location you care to think of apart from deep, dry shade, this may well be a much better regarded plant.

R. repens (Figures 8.52 through 8.56) is native to Eurasia and is recorded as now being present throughout all North America. It is fully hardy (UK H6), deciduous to semi-evergreen and spreads primarily by creeping stolons. It is also spread by seed that although frequently observed to rot, may persist in the soil for 20 years or more. Wood pigeons are known to eat the seeds. Individual plants generally last only a year or two, surviving as rooted crowns over winter before dwindling away the following season after blooming, but they are usually perpetuated by daughter plants produced on vigorous and often far-creeping stems. The number of daughter plants produced by a mother plant is variable and can be anywhere between 1 and 16 or more new clones. This variation is thought to be a density-dependent response to overcrowding in the immediate vicinity of the mother plant.

It has variably incised dark-green leaves with pale patches, which are sometimes clearly and attractively marked. They are divided into three-toothed leaflets that are somewhat hairy, as are the stems. The plant contains small quantities of a toxin that causes salivation, diarrhoea and stomach ache when eaten by most mammals, and deer and rabbits avoid it. Surprisingly, many birds will eat the leaves, particularly geese and chickens.

Ranunculus repens produces panicles of bright, glossy yellow, five-petalled flowers from April to August. The plants are hermaphroditic and are pollinated by bees, butterflies, flies, bugs and beetles. It grows particularly well in semi-shade on moist soil. The ideal soil is a fertile loam that is a neutral to mildly acidic, although it is found growing on heavy clay and moist sandy soil. Plants grow most vigorously on rich, moist and deep soils.

R. repens is often accused of being potassium hungry, and it is also reported to be allelopathic and therefore chemically detrimental to neighbouring plants. However, although both assertions are often reported, both are poorly validated and should be viewed accordingly.

The petals of buttercups are glossy on the inside and reflect their yellow colour due to the inside surfaces being coated by special cells which trap two thin layers of air just beneath the surface and act as a mirror. The curved petals reflect and focus both visible and ultraviolet (UV) light towards the centre of the flowers where they influence nectar secretion rate, its subsequent evaporation and concentration, and act to warm up the pollen-producing stamens, which can increase their growth rate and enhance their chance of fertilization. As many pollinators, including bees, have eyes sensitive in the UV region, it is thought that this is how buttercups use their unique UV reflective and glossy flowers to attract insects.

Buttercups have also been used as a crude method to estimate the age of hedgerows and grasslands based on the number of plants found with additional petals, each extra petal being equated to approx. 7 years presence in the location. Extra petals are thought to be somatic mutations that occur over time and are inherited only via vegetative reproduction since they do not occur on seed-raised plants.

One of the most vigorous of plant species used in T-lawns, it should be used with some caution, although a T-lawn without buttercups seems somehow incomplete, and they are likely to turn up, like it or not. When starting a new lawn, only a few scattered individual plants are required, numbers will increase in time. Ideally only useful ornamental cultivars should be used. They tend to be a just a little less vigorous and can be more appealing to those who have learned to despise the distinctive glossy yellow flowers in gardens, although there is nothing to stop the (brave) use of the common wild form.

Unlike grazing, which seems to have little influence on *R. repens* due to its selective nature, it has been noted that mowing does reduce the vigour of the plant overall. Mown patches tend to be less extensive and lower growing than similar patches in fields and open commons. Its vigour is also ameliorated by soils not particularly rich in nutrients, shallow soils and occasionally dry soils. Whatever the situation and soil, it is worth trying a few creeping buttercup cultivars to see how they manage. They can inform you on your lawn conditions, and since they are not small plants, they can be extracted quite easily if they prove to be problematic.

There are several useful cultivars: *R. repens* 'Flore pleno' has tight fully double flowers and is perhaps the most vigorous of the cultivars, especially if growing in consistently moist soil.

Unfortunately, the cultivars cannot be relied upon to produce true to type seed, although 'Buttered Popcorn' has been found to produce some similar offspring from seed. There is also a white-variegated form 'Snowdrift' that may have been lost to cultivation quite recently. If any of your buttercups start behaving like thugs, then treat them appropriately and yank out as many as you deem necessary.

FIGURE 8.52 *R. repens* 'Flore pleno', here growing through *Veronica officianalis*. There is also a fully double *R. acris* with an RHS AGM with almost identical flowers, but it is much taller growing and should be avoided in T-lawns.

FIGURE 8.53 *R. repens* 'Gloria Spale'. A cultivar that has lost the usual yellow floral pigmentation and has clear white, well-spaced petals that sometimes appear star shaped. It has foliage more finely cut than the common wild form and is rarely a nuisance in T-lawns, being much better behaved and generally less vigorous than its cousins.

FIGURE 8.54   *R. repens* 'Buttered Popcorn'. A cultivar from the United States with yellow variegated leaves. Despite reports to the contrary, in all but the richest moist soil it is generally a little less vigorous than most wild forms. The intensity of yellow in the leaves tends to be greatest in full sun, decreasing with the amount of shade. The flowers are the familiar buttercup type, although occasionally they appear more buttery-yellow than most wild forms. The reliably yellow foliage contributes well to the variety of foliage colours in a T-lawn, and it is a valuable addition in this regard.

FIGURE 8.55   A mix of *R. repens* 'Buttered Popcorn and white-flowered 'Gloria Spale'.

FIGURE 8.56 A wild-type *R. repens* showing somatic mutation in the flowers. Here with 12 petals instead of the usual 5, the plant may represent a 50+ year-old clone. Tentatively named 'Robert' after the gardener in whose garden this was a chance find, this particular form was collected in an enclosed domestic garden in Peckham, London.

Stellaria graminea L.

Lesser stitchwort, little starwort.

TYPE: Ground cover. Floral.
REPRODUCTION: Seed. Rhizomes.
LOCATION: Sun.
SOIL: Well-drained. Neutral to acidic.
USE: *** Many.

One of those plants that is widespread and not uncommon but manages to be hardly noticed, perhaps because it usually inhabits hedgerows and woodland edges where its flowers are often hidden by other vegetation, and that its flowers are considered too small to be grown in gardens.

A British and European native, the common name of stitchwort refers to an old herbal remedy that was brewed to alleviate the side pain known as 'a stitch', and now medically referred to as 'exercise-related transient abdominal pain' (ETAP). Its other common name of starwort is evident from its white star-shaped flowers that can appear on a smooth and lax inflorescence containing around 12 or more well-spaced little blooms.

The clear white flowers are around half the size of its rough-feeling cousin greater stitchwort (*S. holostea*) at 0.5–1 cm in diameter, and although it appears on first glance to have 10 narrow petals, it has only 5, but they are deeply incised. It has 10 stamens with pale white filaments and reddish anthers that bear yellow pollen. Each flower lasts for just 3 days; however, new flowers continue to be produced throughout the summer. Hermaphroditic flowers are usual although some may be partially or completely sterile males. They first appear in late May in small numbers and increase to a peak in late June before trailing off towards September. They are chiefly visited by flies.

It is a sprawling-to-erect and short-to-tall creeping plant of variable height, with weak, straggly much-branched, square, smooth stems. Its leaves are thin lanceolate to linear, pointed, and smooth and may be mistaken for grass leaves, giving it the name in North America of grass-leaved stitchwort. It produces a weak white rhizome that is shortly creeping with fibrous roots. It is less common on alkaline soils preferring neutral to mildly acidic soil that is both moisture retentive and well-drained.

In Tapestry Lawns

A delightful plant to have in a T-lawn. It is generally unobtrusive as it weaves its way through the vegetation and produces around a dozen flowers per stem come June. Although sometimes mistaken for a grass, on closer inspection its leafed lax stems prove it is a forb. It is an adaptive plant with the flowers produced at the height reached by the neighbouring vegetation; it can flower at the T-lawn cut height of 4 cm. It creeps and weaves itself in a manner that suggests the name stitchwort may also have associations with needlework and stitching (Figure 8.57).

It does not generally root as it creeps, and its stems will wither come winter. It needs no special management and is unlikely to be a plant that needs to be removed by lawn gardening. It is not a mowing trigger species. Its continuous flowering during the summer months is especially useful as other British natives pass their floral peak and the flower power of the lawn declines.

As it has the smallest flowers of the native T-lawn stalwart species, it has been used as a basic unit of floriferousness, the 'Stellaria', and been useful in quantifying the floriferousness of tapestry lawns. The average size of each species' flower is stated in terms of how many *Stellaria* flowers it would take to cover the same area, allowing for comparisons to be made between species if you are interested in that sort of thing.

FIGURE 8.57   Self-woven through the lawn, stitchwort's starry flowers can appear in great numbers.

Thymus praecox Opiz.

Wild thyme, creeping thyme, mother of thyme.

Synonym: *Thymus serpyllum*.

TYPE: Scented Foliage. Floral.
REPRODUCTION: Seed.
LOCATION: Sun.
SOIL: Well-drained.
USE: ** Some.

A plant with identity issues. Ornamental horticulture, plant nurseries and garden centres tend to generally name it *Thymus serpyllum*, which is the 'Breckland thyme' that in the UK comes from a very small area around Breckland in East Anglia. Actual *Thymus serpyllum* is thought to be rarely grown, despite what the plant labels say. Another name, not often seen, but that refers to the same plant is *Thymus drucei*. More recently, *Thymus praecox* has been the recognised name and is the one used on the static database 'Plantlist' held by Kew and the Missouri Botanical Gardens. However, the go to and highly authoritative 'Flora of the British Isles' by Clive Stace (so authoritative in fact, that it is commonly referred to as simply 'Stace' by British botanists) names it as *Thymus polytrichus* subsp. *britannicus*. Yet another respected authority, on thymes in particular (the author of the *Thyme Handbook 2009*), suggests that the creeping thymes of Britain should all be regarded as *T. serpyllum* rather than *T. polytrichus* subsp. *britannicus*, particularly since there is such a wide variation in the amount of hairiness present in wild thyme populations and this is the classic identifying characteristic.

Even the common name is reputedly variable based on the local presence or absence of *Thymus serpyllum*. 'Mother of thyme' is used if *T. serpyllum* is present and 'creeping thyme' and/or 'wild thyme' being used where *T. serpyllum* is absent; although this local distinction seems highly dubious, since it would require a rather large proportion of the population to be sure of the plant in a way that current botanists are not.

Thymus praecox (Figures 8.58 and 8.59) is, given ideal conditions, a low-growing, mat-forming evergreen sub-shrub with lax or prostrate creeping stems that root internodally as they spread. The rate of spread is relatively slow. It grows to around 10 cm tall with a spread of about 30 cm. It is native and grows all over the British Isles, although it is less common in the southeast of England where *Thymus pulegioides* (large thyme) is more generally prevalent. It is fully hardy to zone (UK) H5, although it may lose its leaves in cold wet winters. Dense inflorescences of small, tubular, two-lipped, deep pink to purple flowers appear in summer (usually from June to September) on erect flowering stems rising to 10 cm tall. The flowers are nectar rich for their size and attractive to bees. They are hermaphrodite and are pollinated by insects. There is no seedpod, and seeds are found at the bottom of the calyx. The seeds are tiny dark-brown nutlets with four seeds from each individual flower on the flowering stem.

Found on generally on light, sandy and medium loamy soils that are well-drained. It grows across a range of soil pH, from mildly acid and neutral through to alkaline soils, doing particularly well on thin, nutrient-poor, lime-rich or chalky soils. It cannot grow well in shade and can tolerate drought conditions.

Although leaves are aromatic, the strength of the scent varies according to season and habitat, and leaves are usually not considered to be for culinary use. Wild thyme is one of the plants on which both the common blue butterfly and large blue butterfly larvae feed.

In Tapestry Lawns

Under whatever name or label you find them, it is the low-growing and creeping thymes that root along their stems that are the only thymes that grow successfully in T-lawns, and then only in well-drained and predominantly sunny lawns. They do seem to just about manage in clay and highly organic soils, but only if the soil is well-drained and winter wet is not an ongoing issue. It is usually drainage that will determine whether thyme will last in the lawn beyond a year or so; access to sufficient direct sun and neighbourhood competition will determine its longer-term position. It does best when associated with other prostrate and low-growing species.

Being a chaemophyte (a low-growing woody subshrub) rather than the usual T-lawn suitable hemicryptophytes, it behaves in a different manner to most T-lawn plants. This is particularly noticeable in winter when unlike the soft stems of most other T-lawn plants, the mown lightly woody stems perpetuate and may be denuded of leaves if the soil remains wet for extended periods. The stems may turn also turn black, and the plant may rot if drainage is poor. It is something of an indicator plant in this regard. If it is suited to its spot in a T-lawn, the plant can remain mostly semi-evergreen over winter and begin new leaf growth as early as February.

T. praecox tends to do best in shallow or gritty soil lawns but is well worth trying if overall or localised drainage is generally good. The scent from trodden or mown foliage is one of the real pleasures of having this in a T-lawn. It can almost make a T-lawn something you wish to stroke, as it releases a notable and familiar herbal scent. With chamomile, pennyroyal and thyme in a lawn, it is possible to understand why our ancestors might have danced upon them from time to time. The combined scent is delightfully heady stuff.

It can look a bit untidy when not in growth due to its thin lightly woody stems, especially in winter, and may aesthetically benefit from a tidy up in spring if some of the stems have died. Insects of all kinds find its nectar irresistible, and its small flowers are frequently buzzing with bees and butterflies whenever the sun is shining; it is also the preferred food plant for some moth and butterfly caterpillars, which if you happen to chance across, should be left to their own devices. It seems birds will likely find them soon enough.

The true species may be difficult to determine since thymes can be quite promiscuous and hybridisation is not uncommon; however, there are creeping cultivars listed as either *T. praecox* or *T. serpyllum* that can be useful, with floral colours including deep pink, white and violet-mauve. There are also the variegated leaf forms 'Doone Valley' and 'Highland Cream'.

FIGURE 8.58   A white-flowered form of *T. praecox*.

FIGURE 8.59   A pale mauve form growing with chamomile, a plant with which it appears to associate with well.

Trifolium pratense L.

Red clover.

TYPE: Ground cover. Floral.
REPRODUCTION: Seed.
LOCATION: Sun.
SOIL: Well-drained.
USE: ** Some. Scattered individuals.

Red clover seems an odd common name for a plant with pink flowers, until you know that our fore-bears did not think of pink as a distinct colour and saw it as simply a pale red. They weren't wrong; there are no pink wavelengths of light.

Trifolium pratense (Figures 8.60 through 8.62) is a native, short-lived perennial forb that depend-ing on the variety can grow to 0.6 m by 0.6 m. It is a very variable species, with many subspecies, and both erect and prostrate varieties. It is hardy to zone (UK) 6. It can vary between short and tall forms and is commonly a tufted, hairy plant that grows from a crown. Its trifoliate leaves often have a white crescent at the centre of their oval leaflets. The reddish purple or pink solitary flowers are 12–15 mm, in dense, globose heads and usually generally appear from May through to September. They are hermaphrodite, self-incompatible and are pollinated by bees and butterflies, although the nectar content of the flowers is generally less than found in its cousin *Trifolium repens*, and bees tend not to work over the flowers for as long.

Whilst easy to tell apart when in flower, white and red clover can be challenging to identify with only the leaves to go on. Red clover leaves are hairy and white clover leaves are toothed. There is a mnemonic learned by student botanists that can be useful – 'Red hair, white teeth'. Stems are also hollow, will extend to a maximum length determined by the variety, and then branch along its length at nodes, rather than creep and continue to extend and adventitiously root in the manner of *T. repens*.

It is an important food plant for the caterpillars of many butterfly and moth species, and like other clovers it develops a symbiotic relationship with soil rhizobium bacteria that form nodules in its roots. Working in conjunction, plant and bacteria fix atmospheric nitrogen, even in nitrogen rich soils. It forms short, extensively branched tap roots found mostly in the top 12 cm of soil and can be found growing even in compacted soils.

It grows on most soil types, from nutritionally poor light sandy soils and richer loams through to heavier clay soils, although it does best on soils that are moist but well-drained, showing a greater tolerance for wetter soils that *T. repens*. Although more shade tolerant than *T. repens*, it grows best in full sun.

T. pratensis is one of the few T-lawn plants that is in the main, rarely clonal and does not generally spread via stolons, rhizomes or adventitious roots. Having said that, there are new developments in the 'Mattenklee' pasture types of red clover that do exhibit a stoloniferous habit, although these are unlikely to find their way into T-lawns any time soon, since the breeding behind these stoloniferous types is aimed at providing mixed grass and clover fodder for grazing animals rather than for use in garden lawnscapes.

Depending on the variety, red clover is generally a semi-vigorous to vigorous plant, and for that reason, it is best to plant it individually or in small groups of three, scattered about the lawn. It is a plant that you should include if you can, since not only are its pinkish flower a welcome colour in T-lawns but it is also generally more robust than white clover and more likely to survive environmental extremes, such as short periods of drought or prolonged wetness. Upright forms should be avoided since they tend to produce mounds of foliage and dominate the area they inhabit, often becoming mowing trigger plants as a result. It is best to obtain prostrate forms from wildflower seed merchants that make a point of noting that their sources are from prostrate forms.

There is a yellow-veined cultivar sometimes available at specialist nurseries *T. pratense* 'Susan Smith'. It is not a strong grower and must be propagated via cuttings. It manages in T-lawns since although it tries to be an erect plant, it usually flops and enough of it remains under the blade cut height for it to perpetuate in an induced decumbent form.

FIGURE 8.60   *T. pratense* cultivar 'Susan Smith' has yellow veins.

FIGURE 8.61   An unnamed cultivar with multi-tonal green leaves.

FIGURE 8.62   An unnamed pale-green cultivar.

Trifolium repens L.

White clover, Dutch clover, ladino clover, shamrock, microclover.

TYPE: Floral. All year ground cover.
REPRODUCTION: Stoloniferous. Seed.
LOCATION: Partial shade. Sun.
SOIL: Rarely dry. Well-drained.
USE: * Few/** Some (Dark Dancer™ + Red-leaved cultivars).

Trifolium repens is a usually prostrate leguminous hardy forb that usually behaves as a perennial, but it is frequently an annual in warm climates and a biennial in cold. It is a highly variable species. Leaves tend to have long petioles (stalks) and predominantly three leaflets that can be a solid dull green or occasionally marked with a white 'V', rarely with dark red flecks.

White clover flowers typically develop once days lengthen to 14 hours although flowers can be found in small numbers virtually all year, mainly between April and November with most flowers between May and July. Flower heads contain many individual fragrant flowers usually white, between 6 and 10 mm long that can produce 3–4 seeds, with around 75–100 seeds per head. Seed germinates at the soil surface and usually develops a taproot to 1 m deep, but it dies at the end of the first year, and secondary roots developing from the outward spreading stolons become the main root system and the plant develops patches (Figure 8.63).

The subsequent roots of white clover are mainly shallow rarely exceeding 20 cm depth. White clover grows best between annual temperatures of 4°C–22°C in generally moist and partially shaded environments; hot dry periods during the summer reduce growth. It is considered fully hardy but may lose its leaves at temperatures much below −4°C. It is reported to tolerate a wide range of soil pH (4.5–8.2), although it is said not to grow well in alkaline soil, and the best pH range is 6.0–7.0 (mildly acidic to neutral). It does better on moist clay or loam soils than on sandy soils.

It may display considerable phenotypic plasticity allowing it to respond to local conditions and also show genotypic variability allowing different genes to respond differentially to environmental variation. It makes important symbiotic associations with nitrogen-fixing bacteria known as

FIGURE 8.63 A patch of white clover in a grass lawn showing how the plant develops outwards via stolons, with floral stems occurring on new growth. Not so long-ago small amounts of clover seed were routinely added to lawn seed mixes to 'green them up'. Here the greening effect can be seen clearly. Although relevant, it is not so much the addition of clover-synthesised nitrogen to the soil as is often supposed, but rather the green of the clover leaves themselves that has the predominant greening effect.

rhizobia. This association is complex but can lead to the release of the plant nutrient nitrogen into a soil, and because of this, there is a common misperception that clover somehow altruistically 'feeds the soil' and simply growing it is sufficient. It is not.

All organisms use the ammonia (NH_3) form of nitrogen to manufacture amino acids, proteins, nucleic acids and other nitrogen-containing components necessary for growth. And although 80% of the atmosphere is nitrogen gas (N_2), it is unusable by most living organisms. The N_2 needs to be converted to the useable ammonia form. This process is mediated via specialised root-dwelling bacteria known as rhizobia that make an association with legumes such as clover. For the nitrogen to be released from the clover-rhizobium association, the rhizobium must die and this occurs when the clover plant or part of it (roots, leaves, stem) is damaged or dies. In lawns this tends to be due to mowing. The mower will damage and remove biomass from the clover above ground and the plant will respond by reducing the amount of root below ground to keep the plant in balance. When it does so, the root-bound rhizobium are made homeless and die. The initial form of nitrogen available from the dead material is organic, via decomposition. Other soil-living bacteria and fungi utilise it and convert it yet again, leading eventually to a form accessible to plants and other microorganisms. Since its origins were partly from the air rather than solely from the soil, it is regarded as a soil nitrogen input.

In Tapestry Lawns

T. repens is a valuable addition to T-lawns. It is both good ground cover and has excellent floral performance with generous quantities of relatively large flowers much visited by bees and other pollinators. However, it grows strongly and if initially included in large amounts can become overly dominant quite quickly. If the lawn includes many non-stalwart plants it is usually best to stick with *T. repens* 'Dark Dancer™' or red-leaved forms, as these are rarely overwhelming.

White clover's ability to modify its growth pattern in response to environmental stimuli will often see all but the largest ladino animal fodder-types respond to mowing. Compared to an unmown plant, a mown white clover is likely to be shorter and more compact, but it is frequent repeated mowing that keeps it that way. The infrequent mowing of T-lawns means it is generally best to use intermediate or smaller growing forms such as the 'Dutch' and 'microclover' types.

Microclovers (*Trifolium repens* var. 'Pirouette' and 'Pipolina') are selected forms of white clover that have smaller leaves and a lower growth habit, particularly when frequently mown. They are reputed to produce fewer flowers than other forms of white clover, but that may be in floral volume rather than flower number, since in T-lawns they readily produced carpets of white flowers.

Microclover creeps and mixes amongst other plants, without forming distinct patches as is typical for the species. Although termed 'micro', in T-lawns due to the infrequent mowing regime it can grow to the same size as common Dutch clover; it is regular mowing that keeps it 'micro' in traditional lawns. Microclovers generally do well in sunny situations, tending to die away in partial and full shade; however, they can be more heat and drought sensitive, particularly on sand soils than the Dutch type. It is worth noting that due to its widely creeping habit that inclusion of microclover in a T-lawn can produce a sward that looks to be predominantly microclover at certain times of the year, particularly during summer (Figure 8.64).

There are over 70 listed white clover varieties and many cultivars. Unless you have a singular preference for the simplicity of the common white and unmarked leaf form, there are other floral and foliage forms to experiment with. Not all white clover flowers are white and not all the leaf patterns are a simple 'V' variegation on a green background (Figures 8.65 and 8.66).

Named cultivars are not available via seed, but *T. repens* 'Dark Dancer™' (Figure 8.68), 'William', 'Dragons Blood', 'Wheatfen', 'Isabella' and 'Green Ice' can be found at some nurseries. 'Dark Dancer™' is frequently a form of white clover with more than one name. It can also be

FIGURE 8.64 Microclover can grow to the same height as 'Dutch'-type cover under a T-lawn-style mowing regime. With its widely creeping habit, it can extensively cover a lawn with a carpet of white flowers in early summer. Although the floral display is impressive, it is usual to have to mow it at least twice to keep the lawn from being swamped under the vigorously growing foliage.

FIGURE 8.65 Flower colour in *T. repens* can vary between a clear white through pink to a rich maroon. T-lawns do not have to be blanketed with white-only clover flower heads.

FIGURE 8.66 Leaf pattern and colouration is likewise variable, often with red feathering in modern cultivars. Both lighter and darker leaf forms exist, and leaflet number can also move beyond the usual three.

found listed as *T. repens* 'pentaphyllum' or 'atropurpureum' or 'quadrifolium' or a combination of these. It is often described as vigorous but seems to be less so when grown in mixed species T-lawns (Figure 8.67).

'Wheatfen' is thought to be a contributory parent to many maroon-flowered white clovers. Such a form has been of note in the past, with a similar dark-leaved and maroon-flowered white clover being hinted at in John Parkinson's 'Theatrum Botanicum' in 1640. An improved form known as

FIGURE 8.67   *T. repens* 'William' has maroon flowers rather than the usual white and a tendency toward variable leaflet number. It is likely a cross between *T. repens* 'Wheatfen' × 'Green Ice' and was found by William Lyall (Manor Nursery, UK) in his lawn. It can be a mowing trigger plant since it produces long petioles (leaf stems).

'Isabella' is also available. Intriguingly a simple green-leaved form also with maroon-red flowers is still recorded as growing wild on a few of the Scilly Isles off the SW coast of the UK and listed as *T. repens* var. *Townsendii*. There are a number of cultivars on the global market. Most of these tend to be restricted to the United States, although the Danish nursery Gartneriet Råhøj also specialises in ornamental *Trifolium*.

The cultivars have a variety of forms and can vary from prostrate and decumbent to reaching around 10 cm tall, which then puts them into the realms of unsuitable ladino-type clover. Mowing will influence growth habit, as will soil compaction and root restriction. Despite a reputation as being 'aggressive' as singular plants, experience suggests that in multi-species T-lawns the cultivars are generally well-behaved in the short term and the wild type can be the stronger form. However, ladino-type forms and cultivars with large and tall growing leaves are best avoided.

Another tantalising aspect of *T. repens* is that its roots may influence water infiltration into a soil by improving the overall macroporosity (air filled pores), and therefore improve overall drainage; something that has been noted of T-lawns compared to grass swards. It is also food for many invertebrates, is a reliable source of nectar for pollinators, and with other clovers may contribute up to 50% of the diet of wood pigeons (*Columba palumbus*) in winter. Occasionally a virus infection in clovers can produce erratic bleached variegation on leaves. This may disappear over time or may spread in patches (Figure 8.69).

FIGURE 8.68   *T. repens* 'Dark Dancer™' showing the colouration and leaflet number that sometimes sees it listed as *T. repens* 'Purpureum Quadrifolium' or 'Pentaphyllum'. In natural conditions, one four-leaflet white clover leaf occurs for approximately every 10,000 three-leaflet leaves – so you must be lucky to find one.

FIGURE 8.69   Virus-induced variegation in *T. repens* in a grass lawn.

Veronica chamaedrys L.

Germander speedwell.

TYPE: Ground cover. Floral.
REPRODUCTION: Stoloniferous. Adventitious roots.
LOCATION: Sun.
SOIL: Well-drained.
USE: *** Many.

Veronica chamaedrys (Figures 8.70 and 8.71) is the commonest British species of speedwell, found widely across the country. It is an evergreen perennial, hardy to zone (UK) 7 and unless supported by other strong-stemmed plants is low growing and sprawling with lightly ascending stems. It spreads vegetatively by prostrate stems, which root at nodes producing a fibrous root system and spreads rapidly in patches with reproduction from seed thought to be comparatively rare.

The numerous hermaphrodite flowers are various shades of sky blue, mauve, or pale violet with darker radiating lines from a white central eye. They appear as racemes on short spikes from March to July, lasting only a day or so and often change colour after pollination that is usually by flies. The petals are so lightly attached that the least jarring causes them to drop. Flowers close at night and during wet weather. Mature fruit is heart shaped. It is distinguished by toothed hairy leaves and two lines of fine white hairs on opposite sides of the stems, thought to deter some insects.

An adaptable plant that will tolerate almost any soil from alkaline to slightly acid, from heavy clay, through to dry and sandy, although it does best in moist well-drained soils. It can grow in semi-shade or full sun and can be found growing wild in open woodlands, hedge banks, grassland, rock outcrops, upland scree, road verges, railway banks and waste ground.

The name 'germander' is thought to be a corruption of the Latin *chamaedrys* and not a reference to a specific European country. The leaves are sometimes attacked by the gall midge *Jaapiella veronicae*, and white galls rather like small hairy white balls can appear at the ends of the shoots.

There is an unnamed golden-leaved form and a strongly variegated cultivar, *V. chamaedrys* 'Pam', that was found in France and is also known by the name 'Miffy Brute'. Both cultivars and pure species seed can be difficult to source.

In Tapestry Lawns

Another tapestry lawn stalwart, *V. chamaedrys*, is a real doer in most T-lawns, although in moist and well-nourished lawns it can become a trigger plant for mowing since it can form self-supporting mounds in spring. Outside its floral period it generally creeps well through lawns, maintains evergreen patches, and responds well to mowing, seemingly having no difficulty in recovering from occasional mowing.

The eye-catching flowers come in large numbers and can create patches of pale gentian blue, though often at a time when a first mowing is looming large in May. The flowers are delicate and attractive and make you wonder why this isn't a more common garden plant. It can be surprisingly uncomfortable to mow off the many flowering spikes, although during its floral period more may appear to replace those lost, but never with quite the same abundance.

After flowering it changes from growing vertically to growing horizontally and provides good ground cover but can seemingly spread continuously and may need some control after two or more years. The fluffy white balls that sometimes appear at the end of shoots due to a gall midge are not problematic and are often mown off; they can actually act to add to the interest. One very young visitor to a T-lawn remarked that it must be lawn pixies growing tails for bunnies.

FIGURE 8.70 *V. chamaedrys* mixes well with ground ivy (*G. hederacea*).

FIGURE 8.71   An unnamed *V. chamaedrys* cultivar with golden leaves. If grown in competition with green-leaved *V. chamaedrys*, it is likely to be outcompeted by the stronger form.

Veronica officinalis L.

Heath speedwell, common speedwell, gypsy weed.

TYPE: Ground cover. Floral.
REPRODUCTION: Seed. Creeping stolons.
LOCATION: Sun.
SOIL: Well-drained.
USE: *** Many.

Like all plants with *'officinalis'* in their name, *Veronica officinalis* was once in use as a culinary or medical herb – in this case both. A type of bitter tea, a cure for earache and an evil-spirit repellent being some of its reputed properties. Its evil-repellent properties are thought to be pre-Christian in origin but may also stem from the genus name *Veronica*, since the name is related to Saint Veronica, who wiped Jesus' face with a cloth on the way to Calvary. Some members of the genus are reputed to have markings that supposedly resemble those left on the cloth.

It is a native to Britain, Europe, and western Asia. It is a low-growing, hardy (UK) H3, generally evergreen, creeping perennial that roots at its nodes, spreading via prostrate to ascending stems up to 0.3 m, although it rarely reaches above 10 cm, the height being achieved primarily by its floral stems. Its leaves may be oval to elliptical and are softly hairy, and its 6–8 mm flowers are a pale lilac blue with darker veins that occur on long-stalked upright racemes. The flowers are hermaphrodite and self-fertile although the plant is usually pollinated by visiting bees and flies. It is a summer flowering plant with the first flowers usually appearing in late May, and given fine weather, they may continue until August.

It is usually found growing in mostly sunny places such as open dry woodlands, well-drained south-facing banks, established dunes, open grasslands and heathland. It is often associated with moderately acidic or nutrient-leached soils, although it can be found across a range of soil types, from sandy through to loamy clays that can vary between moderately acidic to mildly alkaline.

It is the sole foodplant for the (rare in the UK) beetle *Longitarsus longiseta* that consumes both leaves and roots.

In Tapestry Lawns

Should you have botanically aware witches wanting to picnic on your lawn, having this plant in the mix may help ensure they are only the good type. In T-lawns *V. officinalis* (Figure 8.72) is a discrete plant even when in flower. Its floral stems, although often plentiful and full of flowers, are slender and the flowers tend to be quite pale. It is not a statement plant.

In the rich plant community of a T-lawn, this speedwell can struggle if its leaves are consistently covered by other plants and it appears to be reliant on mowing for its continued presence. Where the frequency of mowing does keep the vegetation in check it can maintain a presence amongst some unlikely neighbours, although it is observed to creep away to less competitive parts of the lawn if given the chance.

Nutrient-poor and regularly dryish lawns where the growth of other plants is compromised by the conditions tend to see it establish well, and it can thrive where other species appear to struggle. It is seen to associate well with other small-leaved prostrate plants and with *Thymus praecox*, *Argentina anserina*, *Potentilla reptans* and low-growing forms of *Chamaemelum nobile*. It does better in sunny spots usually found towards the edges of lawns rather than in its centre, although it manages most spots that are not continually moist and that allow its leaves access to good light.

The leaves are usefully evergreen, with only very hard frosts causing them to wither. Evergreen plants are particularly useful in winter lawns when the completely herbaceous plants such as *Argentina anserina* will disappear and leave patches of soil visible. A notable feature of the leaves is that they respond to low temperatures by changing colour, darkening considerably and producing a multi-tonal effect in winter lawns (Figure 8.73).

FIGURE 8.72   *V. officinalis* amid *Ranunculus repens* 'Flore-pleno'.

FIGURE 8.73   Winter-darkened leaves of *V. officinalis*.

Viola banksii K.R. Thiele & Prober

Australian violet, ivy-leaved violet, Tasmanian trailing violet.

TYPE: Ground cover. Floral.
REPRODUCTION: Stolons.
LOCATION: Sun. Partial shade. Bright shade.
SOIL: Well-drained. Moist.
USE: *** Many.

Usually sold under the name *Viola hederacea* this is yet another of those plants with a question as to its true identity. It seems highly likely that although part of the *Viola hederacea* species complex that the plant found for sale in ornamental horticulture is in fact the closely related species *Viola banksii*. *Viola banksii* (Figure 8.74) has the orbicular leaves and clearly marked flowers found on plants in both British and European garden centres and nurseries. *Viola hederacea* is a less spectacular plant with a more open, less robust habit, smaller leaves and variable but generally less strikingly marked flowers often paler than the form usually seen in garden centres. It is also thought to be a variably hairy species; however, the amount of hairiness appears to be of limited value in identification.

Viola banksii is unique to southeastern Australia, where it has a wide distribution from parts of South Australia and Queensland, down to Tasmania, with at least seven identified forms. The form usually found in ornamental horticulture is described here.

Viola banksii is a prostrate, stoloniferous perennial forb, that on its own can form large, clonal colonies, spreading by layering stems. It is generally hardy UK (H4) and evergreen, although the leaves are not reliably frost hardy and the plant may become herbaceous during winter. The glossy, kidney-shaped leaves are mostly borne in false whorls on contracted stems at ground level. The unscented violet-purple flowers are edged with white and centred by a small, bird-like orange beak. Flowers appear singly on stems up to 15 cm high and occur throughout the summer and well into autumn if the weather remains warm.

It grows best in generally moist locations in sun or semi-shade. In full sun the leaves lose some of their glossy green appearance. In its native habitat it can be found occasionally growing in relatively dry woodland shade, but in the British Isles it tends to flower poorly in such conditions and although seen to survive in bright shade it frequently attempts to migrate to brighter and more moist spots. Its wide native distribution suggests it is adaptable to most soils, and no specific soil types are generally associated with it.

It is described variously as both vigorous and non-vigorous; this appears to be down to the usage and method of cultivation. Alone and in a well-nourished and well-watered hanging basket it can trail extensively and makes a good basket plant. In the ground it can be much slower to spread. There are no records of it setting seed in the UK.

In Tapestry Lawns

An eye-catching summer flowering violet with flowers that can lift above the general level of the lawn and look utterly charming. It is a must have if you can get hold of it. It has survived in wet winter lawns that have frozen solid, so its reputation for being hardy is certainly valid within T-lawns. It can stay evergreen if the winter is mild but usually the leaves succumb after a hard frost or two.

Within T-lawns it is observed to behave in one of three ways: (1) it dies back within a season after planting; (2) it does very little, neither completely dying nor spreading far, rather it inhabits a small defined area, particularly in parts of the lawn that are prone to occasional drying out; (3) or it is quick to creep and spread, notably in more moist and untrodden spots. It does not root well in areas that have notable soil compaction and are frequently trodden, so if you enter onto the lawn from a regular direction you might wisely consider planting this in other less-visited parts of the lawn. Competition with other plants, particularly white clover, appears to hold it in check, and it never spreads as far as you might like. It mingles well but is most noticeable in patches and is one of the few plants that might benefit from two plant tiles being planted adjacent to each other.

There are three hard-to-find cultivars: 'Baby Blue' with much smaller, deep-blue flowers, 'White Glory' with white flowers, and 'Monga Magic' with larger mauve flowers.

FIGURE 8.74 *Viola banksii* creeping into a patch of *Ajuga reptans* 'Burgundy Glow'.

Viola odorata L.

Sweet violet.

TYPE: Ground cover. Floral.
REPRODUCTION: Seed. Stolons.
LOCATION: Shade. Partial shade. (Cool) Sunny moist.
SOIL: Rarely dries out.
USE: *** Many.

Viola odorata (Figures 8.75 and 8.76) is a native, evergreen, perennial forb that has the capacity to spread to 0.5 m via running stolons that root as they creep. Sometimes it has short-lived plagiotropic (subsurface) stems, and it can develop into dense patches when grown alone, although it rarely reaches beyond 15 cm in height. Populations in the north of Britain are thought to be largely introduced and are likely to have garden origins rather than be part of a native wild population. The wild-type flowers tend to be slightly smaller and have a slightly weaker fragrance than their selected garden counterparts.

It is hardy to zone (UK) H5 and will often come into leaf in January. With increasingly mild winters flowers it can be produced as early as February and through to April. The flowers have a characteristic irregular shape (violet-like) and are classed as being 'zygomorphic'. Zygomorphic flowers are symmetrical along one line of symmetry. In the case of *Viola* sp., the line of symmetry is vertical, like the spine of a book, that is, cut a *Viola* flower in half vertically and you'll have its two mirror images.

Predominantly the wild form has flowers that are unsurprisingly violet (dark purple) in colour and give the genus its name. Occasionally white and other colour forms may be found and are much sought after by collectors and gardeners. The plant appears in many medieval tapestries and paintings since the flowers are notably fragrant when first encountered and were an essential component of the medieval strewing herbs mentioned earlier in the book.

V. odorata is hermaphrodite and is primarily pollinated by early flying bees. The pollinated flowers form fruit capsules that bend downwards and open close to the ground. It also produces a second type of flower later in the year although these never open. These later flowers are self-fertile (cleistogamous) and produce unpollinated seeds that are essentially seed-form clones of the parent plant. Both pollinated and unpollinated seeds have a small fatty treat (elaiosome) attached to them that ants find irresistible, and they will frequently take the seed back to their nest. Despite usually requiring a period of chilling before seed germination takes place, the environment provided by the nest significantly increases the rate of seedling emergence, and the first leaves tend to be larger in nest-grown plants. Freshly sown seeds germinate best, whilst stored or packet seed can germinate erratically.

Both the flowers and leaves emit a perfume, and the distilled oil from the flowers is still in demand today. According to perfumiers, the violet leaf provides a rare cut grass and sliced cucumber note to fragrances, while the flowers have a particular sweet and powdery scent. An oddity of this renown scent is that along with fragrant oils known as terpenes, a primary scent component is a ketone compound called ionone that temporarily desensitises the receptors in the nose; sniff a sweet violet flower once and you'll enjoy its sweet perfume, sniff it again and it will seem scentless. The renowned double 'Parma violet' used in sweets and fragrances is thought to be a hybrid of *V. odorata* and *V. suavis*, though it's worth knowing that it requires 40 million violet flowers to make 1 kilo of violet flower absolute, should you fancy making a perfume of your own.

V. odorata is a plant of woodland edges and dappled shade, mossy lawns, and humus rich soils. However, it tolerates a wide range of soil types from sandy and loamy soils through to heavy clay that can range from mildly acidic to mildly alkaline limestones. It grows best on a well-drained moist soil, although it will also grow in hot, sunny positions if the soil does not dry out.

It is the larval food plant for several micromoths and particularly fritillary butterflies that include the High Brown, Dark Green, Silver-washed, Pearl-bordered and Small Pearl-bordered, as well as Weaver's fritillary and the rare visiting Queen of Spain. It also comes with its own leaf-rolling gall midge, *Dasineura odoratae*.

The wild species form is usually a rich deep violet; however, there are quite a number of other colours available, with many cultivars to choose from including doubles. *Viola odorata* 'Sulphurea' is a creamy apricot and is usually found with the name 'Irish Elegance', a slightly more orange and rarely available form 'Orange Excitement' is highly scented with pale orange flowers that appear later than the species, blooming from March to June. 'King of Violets' is highly scented with deep violet double flowers. 'Rosea' is a rosy pink form and is thought to be the most fragrant of all cultivars with a particularly sweet aroma, and 'Perle Rose' has unusual pink-red flowers that are produced later than most other cultivars, flowering in March and April. 'Reine des Neiges' or 'Queen of Snow' is pure white on compact plants (though early flowers can have a faint bluish rinse), and it is one of the longest flowering violets with good scent with the characteristic upward facing flowers of one of its parents 'Konigin Charlotte'. 'The Czar' is thought to have the largest dark violet flowers of all.

If you should wish to pick a few flowers from your lawn rather than just admire them you might consider 'Luxonne', as this is a florist's variety with long-stemmed violet-blue flowers and long-stemmed leaves; fortuitously, it is also good ground cover plant. If you are not too fussed and would just like a bit of everything, then well-known seed and plant companies can generally supply *Viola odorata* 'Miracle® Mixed'.

In Tapestry Lawns

One of the first hemicryptophyte flowers to appear in a T-lawn and particularly effective if planted in large numbers since the lawn will have a low aspect early in the year, and the flowers are usually easily seen. The caveat to that is the amount of nutrition the plant is able to access. Fertile lawns may see the plant develop mounds of foliage rather than presenting their flowers, which will tend to be hidden by the leaves; another very good reason for low soil fertility in T-lawns. Early waking pollinators are drawn to the flowers by scent, bees in particular.

FIGURE 8.75 A mix of *V. 'Rosea'* and two other colour forms mingling with the first daisies.

Usually being the first forb flowers of the season (along with some daisies), it helps to have plenty of them. It is almost impossible to have too many violets in flower in the lawn at the end of winter and early spring as a harbinger of things to come. Fortunately, as the season moves on other plants will overtop them and take their place in the lawns' floral progress. They take mowing well, regenerating with generally smaller and lower-growing leaves and seem happy to take a back seat after their floral period has passed.

The gall midge that causes rolled and swollen leaves does seem to take a liking to some plants more than others and may be a perennial nuisance. However, the repeated defoliation appears to keep the problem in check, although it does not resolve it completely. Removing the midge's favourite plants has also been helpful in lawns where it has occurred in repeatedly noticeable numbers.

FIGURE 8.76 A clump of *V. odorata* 'The Czar' and 'Reine des Neiges'.

9 Other Useful Plant Species

Other plant species can be used in T-lawns; however, unlike the previous list of stalwarts that can generally tolerate a broad range of conditions, these species adapt to lawn life with varying degrees of success.

The realised niche that these species manage to eventually occupy in lawn conditions is notably very variable; they tend to be a bit more specialist in their requirements and in some cases even downright pernickety. The degree of competition with the more generalist stalwart species, particularly white clover, is also an observed influence on their spread and survival. Some plants you may wish to include simply might not suit your lawn mix or location.

The plants listed here have all been used successfully as perennial lawn plants, but it may take some lawn gardening and perhaps slightly more variable mowing regimes to maintain some of them in the long term.

Acaena caesiglauca Bergmans.

Glaucous bidi-bidi.

From New Zealand. In the manner of *Argentina anserina* it is useful in T-lawns for its silver-blue leaves, although this creeping species has many small and lifted white flower-heads that turn into sometimes spiny burrs. It prefers regularly moist but not wet soil in a very well-drained and predominantly sunny site. It does best with other low-growing and ground-hugging species, particularly in poor, shallow or gritty soils.

Acaena microphylla Hook.f.

Scarlet bidi-bidi.

From New Zealand. Similar in appearance to *Acaena inermis* 'Purpurea' with its variably dark chocolate-maroon leaves, this species has eye-catching red burrs and does best in well-drained sunny lawns when mixed with other low-growing species. The timing of its floral period will often see the burrs mown off before they fully ripen, and it is best used as a foliage plant. If some burrs should survive the mower, it is worth noting they are of the hooked-spine form and may be spread farther than the T-lawn by both people and animals.

Ajuga genevensis × *reptans.*

Hybrid bugle.

Of garden origin and a hybrid between the larger non-stoloniferous European *Ajuga genevensis* (upright bugle) and stoloniferous native *Ajuga reptans* (common bugle). This stoloniferous and altogether larger plant than the common bugle shows hybrid vigour and tends to spread farther and grow taller than both its parents with impressive spires of tiered blue flowers in spring. Its stolons tend to be far creeping and create loose open patches that can be surprisingly mobile over time. Mowing affects it more drastically than *A. reptans*; however, it is robust and can recover well.

Micromeria douglasii Benth.

Indian mint.

With the synonyms *Clinopodium douglasii* and *Satureja douglasii*, this plant can be difficult to track down, but it is well worth it. From northwestern USA and British Columbia, Canada, it is a hardy species grown primarily for its scented leaves that emit a clean, fresh spearmint odour when bruised. It is quickly a favourite for those interested in the scents in T-lawns. The white flowers are not large enough to be of notice, and the plant remains discrete within lawns, creeping erratically on slender rooting stems. Succeeds in lawns that are rarely dry and that particularly have partial or dappled shade. Tends to become overly stressed in full sun.

Fragaria vesca L.

Wild strawberry.

Widespread across the northern hemisphere including the UK and Ireland. Often called the alpine strawberry although it occurs at all altitudes, this is one of the original parents of the cultivated strawberry and has small, fragrant, edible fruits. It is a surprisingly adaptable species that sends runners throughout the lawn looking for places it finds suitable. A plant of sunny woodland edges it does well in most lawns, although the leaves can go crispy at the edges if regularly exposed to extended periods of direct sun. There are untested subspecies in the Americas that are similar.

Galium verum L.

Lady's bedstraw.

Widespread across Europe including the UK through to Asia and Japan. A previously much-used strewing herb due to the smell of freshly mown hay it produces when cut. A delightful T-lawn scent-scape plant. It is a summer-flowering, sprawling and generally low-scrambling plant that produces adventitious roots as it spreads. More usually found in meadows and grasslands it is a variable plant that seemingly adjusts its height to the surrounding vegetation. In mown T-lawns it generally stays low and mingles easily through most other plants. Tolerant of light shade, it tends to do better in sunny lawns and survives short periods of drought well.

Geranium pyrenaicum Burm.f.

Hedgerow cranesbill.

From continental Europe, widely naturalised across the UK. A sprawling, generally ever-green and hardy geranium that has the scented leaves you would expect of a hardy gera-nium and can contribute to the T-lawn scent-scape when mown. In flower from April to October its pinkish flowers lift above the foliage on reddish stems. The flowers are not especially large but can be produced in noticeable numbers over a long period. As one floral stem completes flowering, another usually starts generating. It can sprawl largely unnoticed until in flower and is responsive to being mown or walked upon, often becoming a prostrate-type plant as a result of both influences. As its name suggests it has its origins in the Pyrenees and does best in well-drained soil in sunny lawns. Is most noticeable when planted in patches since it does not spread clonally. The form *G. pyrenaicum* f. *albiflorum*, sometimes known as 'Summer Snow' has white flowers, 'Bill Wallis' (shown) has indigo flowers, 'Sarah' is mauve-pink with an indistinct white centre and 'Isparta' (from Turkey) is similar but has larger flowers.

Geranium thunbergii Siebold ex Lindl. & Paxton.

Thunberg's geranium, Gennoshouko.

From Japan. Unusually for a hardy geranium this delightful species is both creeping and late-summer flowering. It can spread by producing adventitious roots where stems are in contact with the soil and comes with white, pink and purple floral forms; there is also a variegated variety 'Jester's Jacket'. The flowers tend to be relatively small but are produced over several months well into autumn. In its native Japan it is generally an open field and grassland species and unsurprisingly it does not do well in shade.

Herniaria glabra L.

Smooth rupturewort.

A British and European native that has been used as a single species lawn alternative under the name 'Green carpet'. It can be pernickety, behaving either as an annual or a short-lived perennial. Having said that, if it establishes in the lawn it is likely to stay there for a while and along with *Potentilla reptans* is useful in covering those sunny spots not much else seems to like. Not fully drought resistant but certainly drought tolerant it forms a tap root and spreads outward across the soil surface from it, managing full sun well. Flowers are insignificant and require a magnifying glass to see. It provides good prostrate coverage and may change hue with colder temperatures.

Hosackia gracilis Benth.

Harlequin lotus.

From the northwest USA and western Canada, and often listed under its synonym *Lotus formosissimus*, it is a plant of damp ditches and similar locations. Previously placed in the genus *Lotus* and looking very much like a bicoloured *Lotus corniculatus*, this strikingly attractive relative does best in semi-shade and moist soil. Its bicolouring can vary from pale pink and yellow to purple and yellow, with flowers produced from April through to July in British T-lawns. Unlike its sun-loving and drought-tolerant cousin *L. corniculatus*, which produces a deep tap root, it is shallower rooting and can produce both stolons and rhizomes, although it tends to be slow to spread.

Houstonia caerulea L.

Bluets.

From eastern USA and Canada. A delightful spring flowering plant common in lawns across eastern USA and Canada but surprisingly pernickety in British lawns. Has thread-like tufted stems in a basal rosette and bright pale blue to whitish flowers with a yellow eye. Across its home range it will grow across a wide range of lawn soils and environmental conditions, but in the UK seems to restrict itself to generally moist and partially shaded conditions, primarily on acidic soils, disappearing quickly in full sun and dry conditions or on alkaline soils. Despite being clonal via spreading and rooting stems, the plants are not long-lived and the primary method of reproduction is seed; mowing off all the seed heads will see the plants swiftly dwindle. The cultivar 'Millard's Variety' seems to be stronger growing form and has deeper blue flowers; it is also more commonly encountered than the wild form that is rare to find in the UK.

Lobelia chinensis Lour.

Chinese lobelia, Bànbiān lián.

From eastern and southern China. Another sometimes pernickety plant that may or may not stay for long in the lawn. It generally forms slowly spreading clumps via rhizomes or creeping stems in seasonally damp but well-drained lawns and tolerates occasionally dry soils when established. The clumps often fragment each year. It manages partial shade but does best in sunny spots. It stays dormant until a period of warm weather wakes it up and may be slow to show. Flowering in summer its blooms are small, vary in hue from pale to dark mauve and are unpredictable in number; sometimes there are few and sometimes many. More of a plant to add to the species number in the lawn than a floral display plant but it is visited by pollinating insects that seem to appreciate it.

Lobelia oligophylla (Wedd.) Lammers.

Hypsela.

Known commonly by the synonym *Hypsela reniformis* this creeping lobelia is from Chile and neighbouring parts of South America. Surprisingly it is classed as having no ornamental value in its home range. It is a nevertheless pretty alpine mat-forming species of *Lobelia* that prefers damp and even marshy places. It mixes well with other lawn lobelias but tends to struggle with taller-growing plants since it does best in a sunny location and does not last long if shaded out. It will spread slowly in damp sunny places and is worth trying in sunny and moist clay lawns.

Nierembergia repens Ruiz & Pav.

Creeping white cup.

From Chile, Argentina and Uruguay and also described as *Nierembergia rivularis*, it is confusingly listed as two distinct species rather than one. Reputedly associated with water and able to survive occasional submersion, this South American beauty has no difficulty in average and well-drained garden soils. Considered to be a tough plant that spreads via freely branching stems or rhizomes just under the soil, although it is not so tough when in tapestry lawn communities and needs to be planted in some numbers to be particularly noticeable and ensure its presence. The noticeable flowers are relatively large at 3–5 cm across, white with a yellow eye and usually lifted just above the leaves, which can be deciduous or semi-evergreen. It flowers during the summer months in sunny but not sun-baked lawns.

Phyla nodiflora (L.) Greene.

Lippia, turkey tangle, frogfruit.

Found in Brazil and the southwestern United States it is generally considered to be a tropical and semi-tropical ground-hugging perennial. It has thick, decumbent trailing stems up to 0.5 m long that form spreading and tangled mats that root when nodes are in contact with the soil. Clusters of white to lavender-pink flowers appear from May to September. Although classed as unreliably hardy in the UK, it has remained perennial in every T-lawn it has been planted in, surviving to −10°C in lawns that have good coverage and are well-drained. Does best in sunny lawns in spots where it is not overly shaded by other plants; when grown in shade it does not flower well. Shows good drought tolerance once established.

Primula elatior L.

Oxlip.

From Europe including the UK. Oxlip is a rare plant with a restricted range on the UK, occurring mostly in parts of East Anglia. However, the False Oxlip, a hybrid of *P. vulgaris* and *P. veris* (*P. × polyantha* Mill.) and parent to many garden polyantha primroses is much more common and more likely to be encountered. Like other British *Primula* sp. both the true and false oxlip can tolerate T-lawns, although they rarely spread far since the relatively high seed heads are invariably removed by the year's first mowing. It is a plant that needs topping up from time to time. As with *Primula vulgaris* the leaves can be almost completely defoliated by mowing, but the plant recovers well. Not considered a clonal plant, it can take some years for clumps to form and is best planted in groups of three or five or more for best effect. A welcome and pretty flower in spring along with lawn bulbs and well worth trying along with the false form too. Prefers moist, well-drained and sunny lawns.

Primula × pruhoniciana.

Primula 'Wanda'.

A magenta-flowered, early-blooming British-bred hybrid between the common primrose *P. vulgaris* and the creeping rootstock alpine *Primula juliae* from the eastern Caucasus mountains and Azerbaijan. Behaves much like native primroses in T-lawns but tends to be neater and less affected by mowing since it does not grow as tall. It is the principal component of the 'Wanda Group' primroses although there are now a variety of different colours available. The relatively dark foliage is retained throughout the year, and the unusual floral colour that can occur from March to April is a good addition to spring lawns. It prefers cool roots and takes summer shade well. Although not listed in the stalwarts section, it maybe the best ornamental-type primrose for T-lawns, although in warm spring weather slugs may sometimes munch on the flowers.

Primula veris L.

Cowslip.

A European and British native. Once recorded to be as common as buttercups across the UK, it suffered a massive decline during the twentieth century and although still widespread is relatively uncommon. Only now is the species showing some early signs of recovery as broadleaf herbicide use is cut back on verges and banksides. Like all British primroses it has a rosette of leaves, but these unroll and claim space in a manner unlike *Primula vulgaris*. It is less susceptible to being defoliated by mowing than its cousin, as the leaves tend to lay much closer to the ground. It has tube-like, rich yellow flowers clustered together at the ends of upright stems; it is these that tend to be removed by mowing and the species must be maintained via additional plants as required. Its upright floral stems in April and May look most effective in groups and scattered clumps, and it may make colonies. It is not a woodland plant and can be planted in sunny locations with good drainage. However, be prepared to chop those flowers; it is often in flower during RHS Chelsea week.

Prunella laciniata (L.) L.

Cut-leaved selfheal, cutleaf selfheal.

From continental Europe where it has a wide range. Potentially native but found rarely growing wild in the UK this is a relative of the native *Prunella vulgaris*. Like its cousin it is perennial but spreads slowly by a buried woody rhizome. It is a larger plant than *P. vulgaris* with narrow incised leaves, and larger more visually impressive flowers (although fewer in number), that are usually white but may also be pink or purple (shown). It will hybridise with *P. vulgaris* and produce intermediate forms. It generally grows taller and appears to be less plastic in its response to mowing than *P. vulgaris*.

Prunella grandiflora (L.) Scholler.

Large-flowered selfheal.

From continental Europe where it has a wide range. The flowers and plant can be twice as large as *Prunella vulgaris* and be so noticeable that the plant is often used in commercial ornamental horticulture where it is sold under the 'Freelander' and 'Loveliness' series names. It is a sprawling, mat-forming, semi-evergreen perennial that spreads by stolons and rhizomes. Its flowers vary from imperial purple through to lavender, pink and white. It is found growing in both meadows and woodlands in Europe and is adaptable to most situations in T-lawns. It is not on the stalwart list simply because it can be short-lived, is of variable height and its response to being mown can also be quite variable. After mowing, the remaining stems can become woody or the plant may die.

Sagina subulata var. *glabrata* (Swartz) C.Presl.

Heath pearlwort, Irish/Scottish moss.

From Europe including the UK. A mat-forming, low-growing, moss-like perennial that has been used as a lawn on its own, particularly the golden-leaved form 'Aurea' that is commonly called 'Scottish moss'. 'Irish moss' is the green form. It is rather unpredictable in how it behaves in T-lawns. It has initially grown and spread well in both partially shady and in sunny lawns, although it appears to respond negatively to unusual periods of wetness or very dry conditions. It does well in sandy loams; however, it has also died or become scarce in the same sandy loam lawns it has previously thrived in, possibly due to unusual weather conditions. It therefore can be regarded as a highly variable lawn resident and may be replaced by its more robust cousin *Sagina procumbens*, which is regarded by many a greenkeeper as a common lawn weed and therefore probably more suitable. It does best when not surrounded by taller competition, is a bit of a loner, but can produce a mass of tiny white flowers that later turn into small white seed pods that resemble pearls, thus its common name.

Soleirolia soleirolii (Req.) Dandy.

Mind your own business.

From Mediterranean Europe. A low-growing prostrate plant for damp and shaded lawns that has become naturalised in most of central and southern England and lowland Wales. Does not grow well in sunny or dry spots and can be frost sensitive although it usually recovers. A small-leaved and ground-hugging creeper that does not take taller competition well but can persist and fill the spaces in parts of the lawn that remain damp and shady where nothing else seems to thrive. Golden ('Aurea') and silvery variegated ('Variegata') forms exist to experiment with.

Taraxacum pseudoroseum Schischk.

Pink dandelion.

From the Tien Shan Mountains of China and central Asia, a dandelion for the lawn. A bico-loured dandelion that behaves just as you would expect a dandelion in the lawn to behave. Pollinator-attracting common dandelions are likely to turn up in T-lawns just as they do in traditional grass lawns, but you can get in early with this one and choose your lawn's dandelions in advance. The pinkness of the flowers is variable both in the proportion of bicolouring and its depth of colour when grown from seed, but if you have a mind to, you can dig up the tap root of your favourite flower form, chop it into chunks and then bury the chunks just below the surface of the lawn in spots you would like to see it. It is certainly a talking point.

Trifolium fragiferum L.

Strawberry clover.

From Europe including the UK, but more common in the south of Britain. Similar in growth form and habit to white clover but with more compact and pinkish flower heads in summer (July and August), usually after white clover has passed its peak flowering period. Can be taller growing than white clover and produce thicker and longer creeping stems but responds to mowing well. It is called 'strawberry' clover not because of the pinkish flowers but for the unusual post-pollination seed heads. This clover is also salt tolerant and is often found growing well in coastal and seaside locations where other clovers struggle.

Trifolium wormskioldii Lehm.

Springbank clover, cow's clover.

From western North America. Common across a wide range of habitats, a purple-and-white bicoloured mat-forming clover with creeping rhizomes and lightly marked leaves. Very attractive in T-lawns if it can survive competition with other clovers. It has decumbent stems that stay lower growing in sun and lift up in dappled shade. Does best in soils that do not dry out for long and when planted in groups. Well worth including.

Veronica repens Clarion ex DC.

Creeping speedwell.

An evergreen creeping plant that hugs the ground and has small white or sometimes bluish-white speedwell flowers during late spring. As a low ground-hugging creeper, it does better in areas with mostly low-growing plants rather than mowing trigger species or in more frequently mown lawns. It does well in semi-shade or in sunny spots that don't completely dry out. Like most shallow-rooted and ground-hugging species, it can be sensitive to low moisture content in surface soil. There is a golden-leaved cultivar 'Sunshine' that has yet to prove perennial in T-lawns.

Viola riviniana Rchb.

Common Dog Violet. Wood Violet. American dog violet. Labrador violet.

From Europe including the UK. One of those plants with a recent botanical reclassification having previously been called *Viola labradorica*. There is a *Viola labradorica* from north-eastern America including Greenland and Labrador, but it is only rarely available and the plant you are much more likely to encounter is *Viola riviniana* Purpurea Group. Purpurea since the leaves are darkly purple (Figure 9.1).

It is a native, non-creeping, evergreen violet that flowers from April to May and inter-mittently throughout the summer. For an earlier flowering violet that looks much the same you might also try *Viola reichenbachiana*—the early dog violet. Interestingly the purple-leaved form does well in full-sun lawns, especially if from self-sown seed, while the green-leaved form does very well in dappled and bright shade. Both forms handling seasonal dryness surprisingly well. They both look best when planted in noticeable patches and can spread randomly across the lawn due to their seedheads explosive dehiscence. There are forms with blue, bluish-purple and pink flowers worth looking for.

FIGURE 9.1   *Viola riviniana* Purpurea Group.

Viola sororia Willd.

American common blue violet.

From North America. Considered a lawn weed in some parts of the USA and almost inevitably a good T-lawn plant, flowering in late spring and early summer. A spreading violet that makes almost solid rhizomatous mats that protrude from the soil, slowly spread and claim the space it inhabits. If you have difficulty getting violets to grow in your lawn, this is the one to try. It is a robust species and takes mowing well, although it can be sensitive to particularly alkaline soils. It is deciduous and loses its leaves in winter.

Although called the common blue violet in the USA, it has a purplish form 'Rubra' (shown), a pure white form, a white form with speckles of blue sold as 'Freckles' and an attractive white form that has a blue centre with prominent blue veins sold as 'Priceana' or under the name of 'Confederate Violet'.

Viola tricolor L.

Heartsease.

From Europe including the UK. An annual therophyte and not a clonal species. A plant that needs to be topped up annually in T-lawns via lots of scattered seed or better still some plug plants. It simply looks lovely in T-lawns even though it's a bit of a general rule breaker. Surprisingly tolerant of being mown once; add some every year and enjoy them.

Geophytes: Bulbs, Corms & Tubers.

Several bulbs, corms and tubers have been used in T-lawns to good effect. For perennial floral displays it is best to use those species that are able to either avoid the mower or complete most or all of their typical life cycle before being defoliated by it. In practice this means using early-spring-flowering species, low-growing summer-flowering species or late-autumn-flowering species.

The geophytes listed here have all been used successfully in T-lawns and are well worth including. Ideally the bulbs, tubers and corms will be added at planting time, but if that should not be possible, they may be added later.

Colchicum autumnale L.

Autumn crocus.

From Europe including the UK. A British native of damp grassy places that appears variously from September to November depending on location. The flowers are followed by up to eight long slender leaves that persist until the first mowing in May. The temptation is to cut the leaves before the first mowing of the year in May to neaten up the look of the winter lawn; however, the winter-long leaves are the price for recurring perennial flowers. Be sure that you don't mind tufts of grass-like leaves in your lawn over winter before you plant it.

Crocus tommasinianus.

Early crocus.

From SW Europe. Amongst the earliest flowering crocus', *Crocus tommasinianus* has long tubed flowers that appear at the same time as the narrow almost grass-like leaves. Blooming can occur in late winter from February to March but may be delayed by extended cold periods; usually the leaves will have almost completely withered by the first mowing in May. It readily produces seed and can spread even in mown tapestry lawns. There are both small and larger types. Useful forms are: 'Albus', 'Barr's Purple', 'Lilac Beauty', 'Roseus', 'Ruby Giant' and 'Whitewell Purple'. Annoyingly, rabbits seem to find both leaves and flowers a tasty snack, and pigeons will peck at the flowers.

Cyclamen coum Mill.

Hardy spring cyclamen.

From the mountains of Turkey, the Caucasus and northern Iran. Although usually sold as corms, cyclamen are not true corms at all and are in fact modified stems and therefore a type of tuber. The often notably variegated leaves appear in autumn and will naturally die back in summer, which is not ideal timing since mowing begins in mid-spring. However, sufficient leaves usually survive beneath the mower's blade for many tubers to be reliably perennial. The flowers may appear in late December or early January and can continue into March. Soils that are well-drained seem to be best in terms of survival; putting a stone or pottery crock under a tuber at planting appears to be helpful in winter wet lawns. Using small tubers has been most effective; they appear to settle into lawn life better than transplanted larger tubers. Perhaps surprisingly, seed has also been a successful method to introduce them to the lawn, although it is important not to remove the first few leaves from the baby tubers if they are to prosper.

Eranthis hyemalis (L.) Salisb.

Winter aconite.

From southern Europe to Turkey. Naturalised in parts of the south and along the east coast of England and Scotland. Usually the first flower of the new year in late January, although the buttercup-like flowers seldom open fully. Properly a plant of woodland and partial to humus-rich and slightly alkaline soils, it can nevertheless be grown in T-lawns and is particularly at home with snowdrops. The larger form, *Eranthis hyemalis* Cilicica Group, usually listed as *E. hyemalis* 'Cilicus', is the best form to use in sunny lawns; the smaller more common form will suit shadier lawns. Best grown in large drifts. The leaves will disappear come May and avoid the mower. The cultivars 'Flore Pleno', 'Guinea Gold' and 'Orange Glow' do not do well in T-lawns and are quick to disappear.

Ficaria verna Huds.

Lesser celandine.

From Europe including the UK and eastern Asia. Still widely referred to by its previous name *Ranunculus ficaria*. Lesser celandine is a tuberous perennial that behaves like a spring ephemeral. It is something of a love/hate plant since its wild form can spread surprisingly quickly and create ground- and plant-covering mats if given the opportunity. It spreads by its many small tubers and also via bulbils and seed. In tapestry lawns it seems less able to successfully create the mats of foliage often seen in grass lawns. However, it is best to avoid the wild form and stick with cultivars; they tend to be less vigorous and some have attractively marked and coloured leaves.

If you wish to provide floral resources for wildlife and do not mind the plant spreading, the single-flowered cultivars are useful. If you would prefer to just enjoy the sprinkling of flowers in the lawn, then it is best to use the double-flowered forms since they almost never produce seed and bulbil production can be less prolific; their rate of spread is generally slower. After flowering, the plant foliage dies back around May and the plants become dormant. There are several named varieties: 'Collarette', 'Brazen Hussy', 'Double Mud', 'Flore Pleno', 'Fried Egg', 'Randall's White' and 'Salmon's White'.

Galanthus nivalis L.

Snowdrop.

From Southern Europe. Widespread across the UK and thought of as a native by many, this familiar spring bulb was not recorded in the wild in the UK until the eighteenth century. Usually thought of as shade-loving plants because they inhabit woodlands, they do however manage to put on a fine early show in open T-lawns. In woodlands the competition is almost non-existent when they are in bloom, but in T-lawns there are plenty of other plants that will be breaking their winter dormancy shortly after the snowdrops have finished flowering. The dappled woodland shade associated with snowdrops may not be available, but partial shade from its immediate neighbours certainly will be. Sometimes in flower in January but more usually in February and March. Use lots for best effect.

Narcissus.

Daffodil.

Narcissus are not easy to recommend for T-lawns since most of them flower in March and April and do not fully complete their bulb recharging before the mower removes their ageing leaves. Aesthetically the taller 'Dutch' types seem too tall for T-lawns and would be trigger mowing plants if they were hemicryptophytes; the dwarfs appear to suit T-lawns better.

Those narcissus that have been included in experimental T-lawns have come back year after year despite the mowing, but they are relatively early-flowering cultivars, and a leaf chop so close to the end of their growing season doesn't appear to do them much harm. *Narcissus* 'Rijnveld's Early Sensation' has been useful, as has 'Peeping Tom' and 'Tête-à-Tête', although both look mangled after the mower has been over them.

Oxalis perdicaria (Molina) Bertero.

Lobate oxalis.

From eastern Argentina, southern Brazil and central Chile. A recent name change for a plant previously known as *Oxalis lobata*. Dormant over winter, fresh green leaves appear in spring and then disappear until the end of summer when they reappear along with golden yellow, honey-scented flowers that can occur from August to November. It has clover-like leaves, although unusually one lobe of the leaflet is raised, thus the common name. It has repeatedly proven to be frost hardy in well-drained soil but is slow to spread in T-lawns and needs to be planted in well-populated groups to be effective.

Oxalis purpurea L.

Grand duchess sorrel, suuring.

From South Africa. In its homeland *Oxalis purpurea* is a sun-loving, summer-deciduous, winter-growing dwarf geophyte that is responsive to rainfall and can be a weed in lawns. However, in the UK with our mild summers and regular rainfall it has not yet decided on how to behave with both leaves and flowers being randomly produced throughout the warmer months, although the flowers tend to be more common in late summer. In T-lawns it forms a small, loose rosette from a small true bulb. The bulbs can spread underground by producing smaller bulbils and patches may form. Each flower has five petals that open like a trumpet during the day and close at night or on overcast days. Flowers are usually pink through to white, with a yellow throat; 'Garnet' has red leaves. It is an exotic-looking addition to late-summer lawns, especially when planted in generous clumps. It is not *Oxalis triangularis* 'Purpurea' or *Oxalis* 'Ken Aslet' (*Oxalis. melanosticta*), which are different species entirely.

Scilla sibirica Haw.

Siberian squill.

From southwestern Russia, the Caucasus and Turkey. Despite its misleading common name, it is not native to Siberia. True blues are rare in T-lawns, and this species in flower en masse in March and April can be very eye-catching. After being pollinated, the leaves and floral stems droop, and after ripening the purple fruits will split and gently deposit their seed nearby. Leaves tend to have completed their useful life cycle by the time the mower is in use. Like early crocuses it easily naturalises with bulb off-shoots and seeds being readily produced. It is a tough plant that, unlike crocuses, rabbits thankfully avoid.

RHS UK Hardiness Ratings

Rating	Temperature ranges °C (°F)	Category
H3	−5 to +1 (23 to 34)	Half-hardy − mild winter
H4	−10 to −5 (14 to 23)	Hardy − average winter
H5	−15 to −10 (5 to 14)	Hardy − cold winter
H6	−20 to −15 (−4 to 5)	Hardy − very cold winter
H7	colder than −20 (<−4)	Very hardy − severe winter

References

1. Klein, E., *A Comprehensive Etymological Dictionary of the English Language*, Vol. 2. 1967, Amsterdam, the Netherlands: Elsevier.
2. Baines, P., *Flax and Linen*, Vol. 133. 1985, London, UK: Osprey Publishing.
3. Pliny, *To Domitus Apollinaris*, in Letters of Pliny, F.C.T. Bosanque, Editor. 2001, The Project Gutenberg: On-line.
4. Fort, T., *The Grass Is Greener*. 2000, London, UK: Harper Collins, p. 23.
5. Harvey, J., *Medieval Gardens*. 1981, London, UK: Batsford.
6. Magnus, A., Alberti Magni De vegetabilibus: libri VII; historiae naturalis pars XVIII. 1867. Reimer.
7. d'Argenville, A.J.D., *La Theorie Et La Pratique Du Jardinage, Oú L'On Traite A Fond Des Beaux Jardins appellés communément Les Jardins De Plaisance Et De Propreté... Avec Les Pratiques de Géométrie nécessaires pour tracer sur le Terrein toutes sortes de figures. Et Un Traité D'Hydrailique Convenable Aux Jardins*. 1747, Paris, France: Chez Pierre-Jean Mariette.
8. Fleming, L. and A. Gore, *The English Garden*. 1979, London, UK: Joseph.
9. Loudon, J.C., *A Treatise on Forming, Improving, and Managing Country Residences: And on the Choice of Situations Appropriate to Every Class of Purchasers*, Vol. 1. 1806, London, UK: Longman, Hurst, Rees, and Orme.
10. Elliot, C., The Alpine Lawn. In *Quarterly Bulletin of the Alpine Garden Society*. 1936, Pershore, Worcestershire: Alpine Garden Society, pp. 373–384.
11. Robbins, P., *Lawn People: How Grasses, Weeds and Chemicals Make Us Who We Are*. 2007, Philadelphia, PA: Temple University Press.
12. Butterfield, B., *Environmental Lawn and Garden Survey*. Burlington, VT: National Gardening Association, 2004.
13. Raciti, S.M., P.M. Groffman, and T.J. Fahey, Nitrogen retention in urban lawns and forests. *Ecological Applications*, 2008. 18(7): pp. 1615–1626.
14. Milesi, C. et al. A Strategy for mapping and modelling the ecological effects of US lawns. In *Joint Symposia URBAN—URS 2005*. 2005, Tempe, AZ: International Society for Photogrammetry and Remote Sensing.
15. Schueler, T.R., *Controlling Urban Runoff: A Practical Manual for Planning and Designing Urban BMPs*. 1987, Washington, DC: Metropolitan Washington Council of Governments.
16. Lush, W., Turf growth and performance evaluation based on turf biomass and tiller density. *Agronomy Journal*, 1990. 82(3): pp. 505–511.
17. Beard, J. and D. Johns, The comparative heat dissipation from three typical urban surfaces: Asphalt, concrete, and a bermudagrass turf. *PR-Texas Agricultural Experiment Station*, 1985. 4329: pp. 59–62.
18. Cook, D.I. and D.F. Van Haverbeke, *Trees and Shrubs for Noise Abatement*. 1971.
19. Mitchell, R. and F. Popham, Greenspace, urbanity and health: Relationships in England. *Journal of Epidemiology and Community Health*, 2007. 61(8): pp. 681–683.
20. Nassauer, J.I., Messy ecosystems, orderly frames. *Landscape Journal*, 1995. 14(2): pp. 161–170.
21. Nassauer, J.I., Z. Wang, and E. Dayrell, What will the neighbours think? Cultural norms and ecological design. *Landscape and Urban Planning*, 2009. 92: pp. 282–292.
22. Rainer, T. and C. West, *Planting in a Post-wild World: Designing Plant Communities for Resilient Landscapes*. 2015, Portland, OR: Timber Press.
23. Del Tredici, P., Nature abhors a garden. *Pacific Horticulture*, 2001. 62(3): pp. 5–6.
24. Raunkiaer, C., *The Life Forms of Plants and Statistical Plant Geography; Being the Collected Papers of C. Raunkiaer*. 1934, Oxford, UK: Clarendon Press.
25. Hautier, Y., P.A. Niklaus, and A. Hector, Competition for light causes plant biodiversity loss after eutrophication. *Science*, 2009. 324(5927): pp. 636–638.
26. Smith, L.S. and M.D.E. Fellowes, The grass-free lawn: Management and species choice for optimum ground cover and plant diversity. *Urban Forestry & Urban Greening*, 2014. 13: pp. 433–442.
27. Doust, L.L., Population dynamics and local specialization in a clonal perennial (*Ranunculus repens*): I. The dynamics of ramets in contrasting habitats. *The Journal of Ecology*, 1981. 69: pp. 743–755.
28. Schaffers, A.P., Soil, biomass, and management of semi-natural vegetation–Part II. Factors controlling species diversity. *Plant Ecology*, 2002. 158(2): pp. 247–268.

29. Connell, J.H., Diversity in tropical rain forests and coral reefs. *Science*, 1978. 199(4335): pp. 1302–1310.

30. Grime, J.P., *The CSR Model of Primary Plant Strategies—Origins, Implications and Tests, in Plant Evolutionary Biology*. 1988, Dordrecht, the Netherlands: Springer. pp. 371–393.

31. Grime, J.P., J.G. Hodgson, and R. Hunt, *Comparative Plant Ecology: A Functional Approach to Common British Species*. 2014: Springer.

32. Fuller, R.A. et al., Psychological benefits of greenspace increase with biodiversity. *Biology Letters*, 2007. 3(4): pp. 390–394.

33. Lindemann-Matthies, P. and T. Marty, Does ecological gardening increase species richness and aesthetic quality of a garden? *Biological Conservation*, 2013. 159: pp. 37–44.

34. Mortimer, S., *Review of the Diet and Micro-habitat Values for Wildlife and the Agronomic Potential of Selected Grassland Plant Species*. 2006, Peterborough, UK: English Nature.

35. Nagashima, H. and K. Hikosaka, Plants in a crowded stand regulate their height growth so as to maintain similar heights to neighbours even when they have potential advantages in height growth. *Annals of Botany*, 2011. 108(1): pp. 207–214.

36. Vermeulen, P.J. et al., Height convergence in response to neighbour growth: Genotypic differences in the stoloniferous plant *Potentilla reptans*. *New Phytologist*, 2008. 177(3): pp. 688–697.

37. Hitchmough, J. and Dunnett N., Introduction to naturalistic planting in urban landscapes. In *The Dynamic Landscape: Design, Ecology and Management of Naturalistic Urban Planting*. 2004, pp. 1–22.

38. Hitchmough, J.D., New approaches to ecologically based, designed urban plant communities in Britain: Do these have any relevance in the United States? *Cities and the Environment (CATE)*, 2008. 1(2): p. 10.

39. Dunnett, N. and J. Hitchmough, The dynamic landscape. In *Design, Ecology and Management of Naturalistic Urban Planting*, N. Dunnet and J. Hitchmough, Editors. 2004, London, UK: Taylor & Francis Group.

40. Francis, R.A., J.D. Millington, and M.A. Chadwick, *Urban Landscape Ecology: Science, Policy and Practice*. 2016, London, UK: Routledge.

41. Biedrzycki, M.L. et al., Root exudates mediate kin recognition in plants. *Communicative & Integrative Biology*, 2010. 3(1): pp. 28–35.

42. Gross, K.L., Effects of seed size and growth form on seedling establishment of six monocarpic perennial plants. *The Journal of Ecology*, 1984. 72: pp. 369–387.

43. Smith, L.S. and M.D. Fellowes, The grass-free lawn: Floral performance and management implications. *Urban Forestry & Urban Greening*, 2015. 14(3): pp. 490–499.

44. Allen, W., D. Balmori, and F. Haeg, *Edible Estates: Attack on the Front Lawn*. 2010, New York: Metropolis Books.

45. Salisbury, A. et al., Enhancing gardens as habitats for plant-associated invertebrates: Should we plant native or exotic species? *Biodiversity and Conservation*, 2017. 26(11): pp. 2657–2673.

46. Smith, L.S. et al., Adding ecological value to the urban lawnscape. Insect abundance and diversity in grass-free lawns. *Biodiversity and Conservation*, 2014. 23: pp. 1–16.

47. Hall, D.M. et al., The city as a refuge for insect pollinators. *Conservation Biology*, 2017. 31(1): pp. 24–29.

48. Boodley, J.W. and R. Sheldrake, *Cornell Peat-Lite Mixes for Commercial Growing*. 1972, Ithaca, NY: New York State College of Agriculture and Life Sciences, Cornell University.

49. Borland, J., Armitage's native plants for North American gardens. *Native Plants Journal*, 2006. 7(2): pp. 153–154.

50. Schröder, R. and R. Prasse, Do cultivated varieties of native plants have the ability to outperform their wild relatives? *PloS One*, 2013. 8(8): p. e71066.

51. Williams, N.M. et al., Bees in disturbed habitats use, but do not prefer, alien plants. *Basic and Applied Ecology*, 2011. 12(4): pp. 332–341.

52. Memmott, J. and N.M. Waser, Integration of alien plants into a native flower–pollinator visitation web. *Proceedings of the Royal Society of London. Series B: Biological Sciences*, 2002. 269(1508): pp. 2395–2399.

53. Morandin, L.A. and C. Kremen, Bee preference for native versus exotic plants in restored agricultural hedgerows. *Restoration Ecology*, 2013. 21(1): pp. 26–32.

54. Morales, C.L. and A. Traveset, A meta-analysis of impacts of alien vs. native plants on pollinator visitation and reproductive success of co-flowering native plants. *Ecology Letters*, 2009. 12(7): pp. 716–728.

55. White, A., *From Nursery to Nature: Evaluating Native Herbaceous Flowering Plants versus Native Cultivars for Pollinator Habitat Restoration*. 2016, Dissertation, University of Vermont.

Index

Note: Page numbers in italic refer to figures respectively.

T - #0288 - 270225 - C264 - 254/178/12 - PB - 9780367144036 - Gloss Lamination